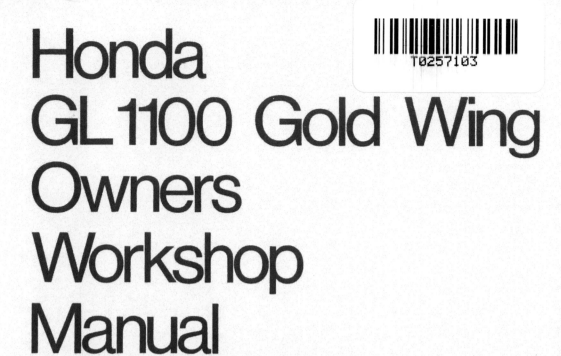

Honda GL1100 Gold Wing Owners Workshop Manual

by Chris Rogers

Models covered:
GL1100 KA. 1085cc. Introduced UK March 1980
GL1100 KB. 1085cc. Introduced UK December 1980
GL1100 DB and DB-DX. 1085cc. Introduced UK January 1981
GL1100 Standard and Interstate. 1085cc. Introduced USA October 1979

ISBN 978 0 85696 669 9

© Haynes Publishing 1993

Printed in the UK   *(669–8S4)*

ABCDE
FGHIJ
KLMNO
PQ
2

Haynes Publishing
Sparkford Nr Yeovil
Somerset BA22 7JJ England

Haynes Publications, Inc
859 Lawrence Drive
Newbury Park
California 91320 USA

Acknowledgements

Our thanks are due to P.R. Taylor and Sons of Chippenham and Calne, Wiltshire, for the loan of the machine featured on the front cover of this manual and in the photographs throughout the text. We should like to thank them also for providing information and much useful advice throughout the project.

Honda (UK) Ltd kindly gave permission to reproduce line drawings; the Avon Rubber Company provided information on tyre fitting and NGK Spark Plugs (UK) Limited furnished advice about sparking plug conditions.

About this manual

The purpose of this manual is to present the owner with a concise and graphic guide which will enable him to tackle any operation from basic routine maintenance to a major overhaul. It has been assumed that any work would be undertaken without the luxury of a well-equipped workshop and a range of manufacturer's service tools.

To this end, the machine featured in the manual was stripped and rebuilt in our own workshop, by a team comprising a mechanic, a photographer and the author. The resulting photographic sequence depicts events as they took place, the hands shown being those of the author and the mechanic.

The use of specialised, and expensive, service tools was avoided unless their use was considered to be essential due to risk of breakage or injury. There is usually some way of improvising a method of removing a stubborn component, provided that a suitable degree of care is exercised.

The author learnt his motorcycle mechanics over a number of years, faced with the same difficulties and using similar facilities to those encountered by most owners. It is hoped that this practical experience can be passed on through the pages of this manual.

Where possible, a well-used example of the machine is chosen for the workshop project, as this highlights any areas which might be particularly prone to giving rise to problems. In this way, any such difficulties are encountered and resolved before the text is written, and the techniques used to deal with them can be incorporated in the relevant sections. Armed with a working knowledge of the machine, the author undertakes a considerable amount of research in order that the maximum amount of data can be included in this manual.

Each Chapter is divided into numbered sections. Within these sections are numbered paragraphs. Cross reference throughout the manual is quite straightforward and logical. When reference is made 'See Section 6.10' it means Section 6, paragraph 10 in the same Chapter. If another Chapter were intended the reference would read, for example, 'See Chapter 2, Section 6.10'. All the photographs are captioned with a section/paragraph number to which they refer and are relevant to the Chapter text adjacent.

Figures (usually line illustrations) appear in a logical but numerical order, within a given Chapter. Fig. 1.1 therefore refers to the first figure in Chapter 1.

Left-hand and right-hand descriptions of the machines and their components refer to the left and right of a given machine when the rider is seated normally.

Motorcycle manufacturers continually make changes to specifications and recommendations, and these, when notified, are incorporated into our manuals at the earliest opportunity.

We take great pride in the accuracy of information given in this manual, but motorcycle manufacturers make alterations and design changes during the production run of a particular motor-cycle of which they do not inform us. No liability can be accepted by the authors or publishers for loss, damage or injury caused by any errors in, or omissions from, the information given.

Contents

Note General description and specifications are given in each Chapter immediately after the list of contents. Fault diagnosis is given at the end of each Chapter

Right-hand view of the 1981 UK Honda GL1100

5

Close-up of the engine and gearbox unit

Introduction to the Honda GL1100 Gold Wing

The present Honda empire, which started in a wooden shack in 1947, now occupies vast factory space, covering all aspects of modern motorcycle design, research, testing and production. The facilities available to the large work force are second to none.

The first motorcycle to be imported into the UK in the early 1960s, the 250 cc twin cylinder 'Dream', was the thin edge of the wedge which has been the Japanese domination of the motorcycle industry. Strange it looked too, to Western eyes, with pressed steel frame, and 'square' styling.

There was, however, nothing strange about the way these modern machines were accepted, and sales were impressive. The trend set by these early machines, that of almost unheard of sophistication, especially on machines of small engine capacity, allied to a remarkably reasonable selling price and a lively performance, ensured their success and started the name Honda on its way to becoming known in virtually every home in the country.

Honda now offer machines to the public to cover every conceivable aspect of motorcycling. These include many four-stroke single cylinder models of both two- and four-valve designs, varying in capacity from 50 cc to 500 cc. Included within the range of on and off road single cylinder machines are variations as diverse as a three-wheeled, go-anywhere, fun machine, the ATC, and the fully equipped XL500S model.

Sports and touring machines of transverse four-cylinder engine design form the mainstay of Honda's medium to upper capacity range, the general trend being broken by a plush 500 cc V-twin engined sports/tourer, the CX 500. Honda's top of the range models also break with this general type of design and are the CBX, a transverse six-cylinder engined machine of 1047 cc, and the subject of this manual, the GL1100 Gold Wing, a flat-four engined machine of 1085 cc.

Designed as the 'ultimate' touring machine and introduced into the USA in October of 1979 and into the UK in March of 1980, the GL1100 is a radically improved version of the discontinued GL1000 Gold Wing. The many improvements made include the adoption of a new suspension system which employs air as a variable pre-load medium; the introduction of a crankshaft mounted CDI ignition system with a vacuum-controlled advance unit; the inclusion of an accelerator pump into one of the four smaller bore carburettors and many changes, both detail and major, to the engine and transmission thus making the unit more efficient overall.

Honda also offer the GL1100 as a ready made 'full dress' tourer complete with fairing, topbox, panniers and crashbars. Known as the Interstate in the USA and, more recently, the GL1100 DX-B Gold Wing Deluxe in the UK, this machine is the first produced by any of the 'big four' Japanese manufacturers to be marketed as a complete touring package. At the time of publication comprehensive information for the GL1100 DX-B (UK) was not available. In general, however, the general procedures are similar to those for the Interstate model, and in many cases specifications will be as for the Standard UK model.

Model dimensions and weights

	Standard	Interstate
Overall length	2345 mm (92.3 in)	2405 mm (94.7 in)
Overall width	920 mm (36.2 in)	920 mm (36.2 in)
Overall height	1195 mm (47.0 in)	1500 mm (59.1 in)
Wheelbase	1605 mm (63.2 in)	1605 mm (63.2 in)
Ground clearance	145 mm (5.7 in)	145 mm (5.7 in)
Dry weight	267 kg (589 lb)	305 kg (672 lb)
Wet weight	290 kg (639 lb)	336 kg (740 lb)
Gross vehicle weight rating	501 kg (1105 lb)	501 kg (1105 lb)

Ordering spare parts

When ordering spare parts for any Honda, it is advisable to deal direct with an official Honda agent, who should be able to supply most items ex-stock. Parts cannot be obtained from Honda (UK) Limited direct; all orders must be routed via an approved agent, even if the parts required are not held in stock.

Always quote the engine and frame numbers in full, and colour when painted parts are required.

The frame number is located on the side of the steering head, and the engine number is stamped on the crankcase immediately to the rear of the oil pressure warning switch.

Use only parts of genuine Honda manufacture. Pattern parts are available, some of which originate from Japan, but in many instances they may have an adverse effect on performance and/or reliability. Further, the use of non-original parts may invalidate the warranty conditions laid down by Honda. Honda do not operate a 'service exchange' scheme.

Some of the more expendable parts such as sparking plugs, bulbs, tyres, oils and greases etc., can be obtained from accessory shops and motor factors, who have convenient opening hours, and can often be found not far from home. It is also possible to obtain parts on a Mail Order basis from a number of specialists who advertise regularly in the motorcycle magazines.

Location of frame number

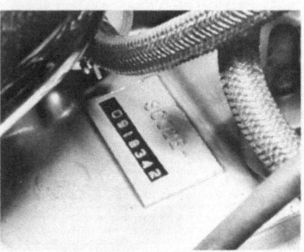

Location of engine number

Safety first!

Professional motor mechanics are trained in safe working procedures. However enthusiastic you may be about getting on with the job in hand, do take the time to ensure that your safety is not put at risk. A moment's lack of attention can result in an accident, as can failure to observe certain elementary precautions.

There will always be new ways of having accidents, and the following points do not pretend to be a comprehensive list of all dangers; they are intended rather to make you aware of the risks and to encourage a safety-conscious approach to all work you carry out on your vehicle.

Essential DOs and DON'Ts

DON'T start the engine without first ascertaining that the transmission is in neutral.

DON'T suddenly remove the filler cap from a hot cooling system – cover it with a cloth and release the pressure gradually first, or you may get scalded by escaping coolant.

DON'T attempt to drain oil until you are sure it has cooled sufficiently to avoid scalding you.

DON'T grasp any part of the engine, exhaust or silencer without first ascertaining that it is sufficiently cool to avoid burning you.

DON'T allow brake fluid or antifreeze to contact the machine's paintwork or plastic components.

DON'T syphon toxic liquids such as fuel, brake fluid or antifreeze by mouth, or allow them to remain on your skin.

DON'T inhale dust – it may be injurious to health (see *Asbestos* heading).

DON'T allow any spilt oil or grease to remain on the floor – wipe it up straight away, before someone slips on it.

DON'T use ill-fitting spanners or other tools which may slip and cause injury.

DON'T attempt to lift a heavy component which may be beyond your capability – get assistance.

DON'T rush to finish a job, or take unverified short cuts.

DON'T allow children or animals in or around an unattended vehicle.

DON'T inflate a tyre to a pressure above the recommended maximum. Apart from overstressing the carcase and wheel rim, in extreme cases the tyre may blow off forcibly.

DO ensure that the machine is supported securely at all times. This is especially important when the machine is blocked up to aid wheel or fork removal.

DO take care when attempting to slacken a stubborn nut or bolt. It is generally better to pull on a spanner, rather than push, so that if slippage occurs you fall away from the machine rather than on to it.

DO wear eye protection when using power tools such as drill, sander, bench grinder etc.

DO use a barrier cream on your hands prior to undertaking dirty jobs – it will protect your skin from infection as well as making the dirt easier to remove afterwards; but make sure your hands aren't left slippery. Note that long-term contact with used engine oil can be a health hazard.

DO keep loose clothing (cuffs, tie etc) and long hair well out of the way of moving mechanical parts.

DO remove rings, wristwatch etc, before working on the vehicle – especially the electrical system.

DO keep your work area tidy – it is only too easy to fall over articles left lying around.

DO exercise caution when compressing springs for removal or installation. Ensure that the tension is applied and released in a controlled manner, using suitable tools which preclude the possibility of the spring escaping violently.

DO ensure that any lifting tackle used has a safe working load rating adequate for the job.

DO get someone to check periodically that all is well, when working alone on the vehicle.

DO carry out work in a logical sequence and check that everything is correctly assembled and tightened afterwards.

DO remember that your vehicle's safety affects that of yourself and others. If in doubt on any point, get specialist advice.

IF, in spite of following these precautions, you are unfortunate enough to injure yourself, seek medical attention as soon as possible.

Asbestos

Certain friction, insulating, sealing, and other products – such as brake linings, clutch linings, gaskets, etc – contain asbestos. *Extreme care must be taken to avoid inhalation of dust from such products since it is hazardous to health.* If in doubt, assume that they *do* contain asbestos.

Fire

Remember at all times that petrol (gasoline) is highly flammable. Never smoke, or have any kind of naked flame around, when working on the vehicle. But the risk does not end there – a spark caused by an electrical short-circuit, by two metal surfaces contacting each other, by careless use of tools, or even by static electricity built up in your body under certain conditions, can ignite petrol vapour, which in a confined space is highly explosive.

Always disconnect the battery earth (ground) terminal before working on any part of the fuel or electrical system, and never risk spilling fuel on to a hot engine or exhaust.

It is recommended that a fire extinguisher of a type suitable for fuel and electrical fires is kept handy in the garage or workplace at all times. Never try to extinguish a fuel or electrical fire with water.

Note: *Any reference to a 'torch' appearing in this manual should always be taken to mean a hand-held battery-operated electric lamp or flashlight. It does **not** mean a welding/gas torch or blowlamp.*

Fumes

Certain fumes are highly toxic and can quickly cause unconsciousness and even death if inhaled to any extent. Petrol (gasoline) vapour comes into this category, as do the vapours from certain solvents such as trichloroethylene. Any draining or pouring of such volatile fluids should be done in a well ventilated area.

When using cleaning fluids and solvents, read the instructions carefully. Never use materials from unmarked containers – they may give off poisonous vapours.

Never run the engine of a motor vehicle in an enclosed space such as a garage. Exhaust fumes contain carbon monoxide which is extremely poisonous; if you need to run the engine, always do so in the open air or at least have the rear of the vehicle outside the workplace.

The battery

Never cause a spark, or allow a naked light, near the vehicle's battery. It will normally be giving off a certain amount of hydrogen gas, which is highly explosive.

Always disconnect the battery earth (ground) terminal before working on the fuel or electrical systems.

If possible, loosen the filler plugs or cover when charging the battery from an external source. Do not charge at an excessive rate or the battery may burst.

Take care when topping up and when carrying the battery. The acid electrolyte, even when diluted, is very corrosive and should not be allowed to contact the eyes or skin.

If you ever need to prepare electrolyte yourself, always add the acid slowly to the water, and never the other way round. Protect against splashes by wearing rubber gloves and goggles.

Mains electricity and electrical equipment

When using an electric power tool, inspection light etc, always ensure that the appliance is correctly connected to its plug and that, where necessary, it is properly earthed (grounded). Do not use such appliances in damp conditions and, again, beware of creating a spark or applying excessive heat in the vicinity of fuel or fuel vapour. Also ensure that the appliances meet the relevant national safety standards.

Ignition HT voltage

A severe electric shock can result from touching certain parts of the ignition system, such as the HT leads, when the engine is running or being cranked, particularly if components are damp or the insulation is defective. Where an electronic ignition system is fitted, the HT voltage is much higher and could prove fatal.

Routine maintenance

Periodic routine maintenance is essential to keep the motorcycle in a peak and safe condition. Routine maintenance also saves money because it provides the opportunity to detect and remedy a fault before it develops further and causes more damage. Maintenance should be undertaken on either a calendar or mileage basis depending on whichever comes sooner. The period between maintenance tasks serves only as a guide since there are many variables eg; age of machine, riding technique and adverse conditions.

The maintenance instructions are generally those recommended by the manufacturer but are supplemented by additional tasks which, through practical experience, the author recommends should be carried out at the intervals suggested. The additional tasks are primarily of a preventative nature, which will assist in eliminating unexpected failure of a component or system, due to wear and tear, and increase safety margins when riding.

All the maintenance tasks are described in detail together with the procedures required for accomplishing them. If necessary, more general information on each topic can be found in the relevant Chapter within the main text.

Although no special tools are required for routine maintenance, a good selection of general workshop tools is essential. Included in the tools must be a range of metric ring or combination spanners, a selection of crosshead screwdrivers, and two pairs of circlip pliers, one external opening and the other internal opening. Additionally, owing to the extreme tightness of most casing screws on Japanese machines, an impact screwdriver, together with a choice of large or small cross-head screw bits, is absolutely indispensable. This is particularly so if the engine has not been dismantled since leaving the factory.

Weekly, or every 200 miles (320 km)

---

1 Tyres

Check the tyre pressures. Always check the pressure when the tyres are cold as the heat generated when the machine has been ridden can increase the pressures by as much as 8 psi, giving a totally inaccurate reading. Variations in pressure of as little as 2 psi may alter certain handling characteristics. It is therefore recommended that whatever type of pressure gauge is used, it should be checked occasionally to ensure accurate readings. Do not put absolute faith in 'free air' gauges at garages or petrol stations. They have been known to be in error.

Inspect the tyre treads for cracking or evidence that the outer rubber is leaving the inner cover. Also check the tyre walls for splitting or perishing. Carefully inspect the treads for stones, flints or shrapnel which may have become embedded and be slowly working their way towards the inner surface. Remove such objects with a suitable tool. The thing for getting stones out of horses' hooves is ideal!

2 Battery

Check the electrolyte level in the battery and replenish, if necessary, with distilled water. Do not use tap water as this will reduce the life of the battery. If the battery is removed for filling, note the tracking of the battery breather pipe which should be replaced in the same position, ensuring that the pipe is not kinked or blocked. If the breather pipe is restricted and the battery overheats for any reason, the pressure produced may, in extreme cases, cause the battery case to fail and a liberal amount of sulphuric acid to be deposited on the electrical harness and frame parts.

3 Engine oil

Check the engine oil level by viewing through the sight glass in the right-hand crankcase. If necessary, rotate the 'wiper' pad by means of a screwdriver applied to the screw in the centre of the window. This will remove any carbon build-up which may obscure the oil level. Replenish the engine oil so that the level is up to the upper level mark. The correct oil grade is SAE 10W/40.

4 Coolant

Open the flap on top of the dummy fuel tank. With the engine running and at normal running temperature, check the coolant level in the radiator expansion tank. If necessary, remove the tank cap and replenish the coolant up to the 'Full' level line. The coolant comprises water mixed with an ethylene glycol base anti-freeze, in a 50/50 ratio. Ideally, distilled water should be used in the cooling system, to reduce corrosion and 'furring-up'. Tap water can be used, particularly if it is known to be soft. For ease of maintenance, a supply of coolant mixture can be made up and used to replenish the system, as necessary. Do not allow the anti-freeze constituent to fall below 40% as the anti-corrosion properties of the coolant mixture will be reduced below an acceptable level. Ensure also that the anti-freeze used is of a type acceptable for use with an aluminium engine. **Warning:** Do not remove the radiator cap when the engine is hot as the reduction in pressure will cause the coolant to boil and exit through the filler orifice.

5 Brake fluid

Check the hydraulic fluid level in the front brake master cylinder reservoir. Before removing the reservoir cap and diaphragm place the handlebars in such a position that the reservoir is approximately vertical. This will prevent spillage. The fluid should lie between the upper and lower lines on the reservoir body. Replenish, if necessary, with hydraulic brake fluid of the correct specification, which is DOT 3 (USA) or SAE-J1703. If the level of fluid in either of the reservoirs is excessively low, check the pads for wear. If the pads are not worn, suspect a fluid leakage in the system. This must be rectified immediately. In the case of the rear brake, check the fluid level as described above. The reservoir is located behind the right-hand sidepanel.

Use an accurate gauge to check the tyre pressures

Maintain battery electrolyte level between upper and lower level marks

Check engine oil level through the crankcase sight glass

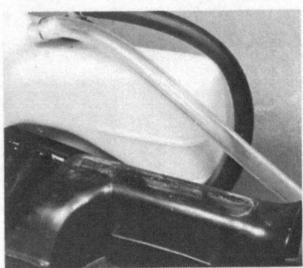

Coolant level should be maintained on FULL line

Check brake fluid level in both front ...

... and rear brake master cylinder reservoirs

6 Electrical system

Check that the various bulbs are functioning properly, paying particular attention to the rear lamp. It is possible that one of the rear lamp or brake lamp filaments has failed but gone unnoticed. Check that the indicators and horns operate normally. Clean all lenses. If any of the fuses has blown recently, check that the source of the problem has been resolved and that the spare fuse has been renewed.

7 Safety inspection

Give the whole machine a close visual inspection, checking for loose nuts and fittings, frayed control cables and damaged brake hoses etc.

8 Legal inspection

As well as inspecting the tyres and lights as detailed in the above Sections, ensure that both the horns and speedometer are working properly.

---

6 monthly, or every 4000 miles (6400 km)

Carry out the tasks listed in the weekly/200 mile maintenance Section and then complete the following:

1 Brake pad wear

To check wear on the front brake pads examine the pads through the small window in the main caliper units. If the red mark on the periphery of any pad has been reached, both pads in that set must be renewed. The rate of wear of the two sets is similar so it is probable that they will require renewal at the same time in any case.

Check the rear brake using a similar procedure to that adopted for the front brakes. Refer to Chapter 6 for details of brake pad renewal.

2 Clutch adjustment

Check the amount of free play at the handlebar lever end. This should be within 10 - 20 mm (0.4 - 0.8 in).

Minor adjustments may be made to obtain this setting by turning the adjuster at the handlebar lever end of the operating cable. Expose the adjuster by pulling back the rubber dust cover, loosen the lock ring and turn the adjuster to obtain the correct amount of free play at the lever end. Tighten the lock ring and relocate the dust cover. It must be noted, however, that it is not permissible to expose more than 12 mm (0.5 in) of adjuster thread beyond the handlebar lever bracket and if it is necessary to do this to obtain the specified amount of free play, then the clutch must be adjusted as follows.

Full clutch adjustment is made by first loosening the lock ring of the adjuster at the handlebar lever end of the operating cable and turning the adjuster fully in. Tighten the lock ring to retain the adjuster in position. Remove the access cap from the clutch cover to expose the adjuster arm screw and locknut. Loosen the locknut and turn the screw until a slight resistance is felt. Note the position of the screw and turn it anti-clockwise one complete turn; hold it in this position and tighten the locknut. Note the condition of the O-ring on the access cap and renew if necessary before refitting and tightening the cap. Move to the adjuster at the engine bracket end of the control cable and loosen the locknut. Turn the adjuster to obtain the required 10 - 20 mm (0.4 - 0.8 in) of free play at the handlebar lever end and tighten the locknut to retain the adjuster in position.

Carry out a final check to ensure all distorted components have been correctly tightened before test riding the machine to check for correct clutch operation.

3 Carburettor adjustment – idle speed

Start the engine and allow it to run until normal operating temperature is reached. Close the throttle and note the engine idle speed; this should be 950 ± 100 rpm. Adjustment may be made by turning the throttle stop screw located immediately below the throttle operating pulley.

Remove cover to inspect brake pads for wear

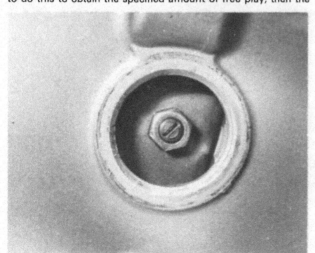

Remove clutch cover access cap to expose adjuster arm screw and locknut

Complete adjustment by turning clutch cable end adjuster

4 Sparking plugs

Remove, clean and adjust the sparking plugs. Carbon and other deposits can be removed, using a wire brush, and emery paper or a file used to clean the electrodes prior to adjusting the gaps. Probably the best method of sparking plug cleaning is by having them shot blasted in a special machine. This type of machine is used by most garages. If the outer electrode of a plug is excessively worn (indicated by a step in the underside) the plug should be renewed. Adjust the points gap on each plug by bending the outer electrode only, so that the gap is within the range 0.6 - 0.7 mm (0.024 - 0.028 in). Before fitting the plugs, smear the threads with graphited grease; this will aid subsequent removal. If replacement plugs are required the correct types are listed in the Specifications Section of Chapter 4.

5 Crankcase breather cleaning

On USA models it is necessary to clean the crankcase breather storage reservoir by detaching it from its frame mounting and inverting it to allow any deposits to drain into a container. Removal of the reservoir is easily achieved by unclipping the feed tube clamp and pulling the tube end off the reservoir stub; the reservoir may then be detached from the frame by unscrewing the single retaining bolt.

If the machine is ridden in a humid climate, in rainy conditions, or at full throttle for long periods of time, then the transparent section of the drain tube should be inspected to check the deposit level at more frequent intervals and the reservoir removed and drained accordingly.

Yearly, or every 8000 miles (12 800 km)

Carry out the operations listed in the previous Sections and then carry out the following:

1 Air filter renewal

To remove the air filter element from the machine, open the dummy fuel tank compartment covers and lift out the tool tray. Unscrew and remove the single wingnut and lift out the air filter cover along with its seal. Remove the air filter element and seal from the filter housing, discard the element and inspect the seal for damage and deterioration before refitting a serviceable item. Fit a new filter element, the filter cover and seal, refit and tighten the wingnut and refit the tool tray before closing the compartment covers.

Do not run the engine with the element removed as the weak mixture caused may result in engine overheating and damage to the cylinders and pistons. A weak mixture can also result if the rubber sealing rings on the element are perished or omitted.

2 Engine oil and filter renewal

Start the engine and allow it to run until normal operating temperature is reached. This will thin the oil and allow it to drain more easily. Place the machine on its centre stand and position a container of more than 4.0 litres (1.25 gallons approx) capacity below the front of the engine. Remove the oil filler cap followed by the oil drain plug, which is located below the filter housing. Allow the oil to drain completely and then remove the oil filter housing, complete with element.

Remove the oil filter element and thoroughly clean the inside of the filter housing. Fit a new oil filter into the filter housing and renew the O-ring which forms the housing to engine front cover seal. Refit the housing to the front of the engine, noting that the housing should be fitted so that the

aligning tabs align either side of the boss on the water pump cover. Tighten the housing bolt fully to a torque setting of 2.7 - 3.3 kgf m (20 - 24 lbf ft). Refit and tighten the engine oil drain plug, to a torque of 3.5 - 4.0 kgf m (25 - 29 lbf ft) after having first renewed the sealing washer.

Replenish the engine through the filler orifice with approximately 3.2 litres (3.4/2.8 US/Imp quarts) of SAE 10W - 40 motor oil until the oil level reaches the upper mark in the sight glass. The machine must be standing on level ground when this check is made.

3 Cooling system

Inspect the radiator core externally for clogging by leaves, flies or road dirt. If required, remove obstructions using a high pressure air hose. Bent fins can be straightened carefully, using a suitable tool such as a screwdriver.

The cooling system is sealed and pressurised and as such if rapid reduction of the coolant level is observed, leakage should be suspected. Check the hoses for cracking or splitting, particularly where the screw clips exert pressure. Check also for leakage at the thermostat feed pipes and the coolant drain plug. If leakage is discovered, it will be necessary to drain the cooling system as follows, before repair can take place. Seepage around a hose at a union may be cured by slight tightening at the relevant hose clip.

Place a clean container of more than 3.4 litres (about 1 gallon) capacity below the water pump housing at the front of the engine. Remove the radiator pressure cap by pressing the cap downwards and rotating in an anti-clockwise direction. Do **not** remove the radiator cap when the engine is still hot. Remove the drain plug from the water pump cover and allow the coolant to drain. The cause of leakage can now be remedied. If seepage has occurred at the thermostat feed pipes, the O-rings should be renewed. Refit and tighten the drain plug after having first renewed its sealing washer. Refill the radiator slowly, to allow as much of the air in the system to escape as possible; about 3.4 litres (3.6/3.0/US/Imp quarts) of coolant will be needed to achieve this.

The cooling system will now have to be bled so that all air is removed. This is accomplished by starting the engine, which should be allowed to run at about 900 rpm for 10 minutes, revving up the engine for the last 30 seconds to accelerate bleeding. The coolant level will fall, indicating the expulsion of air. Top up the radiator and refit the cap. If necessary, replenish the expansion tank so that the level is just below the upper level mark.

Remove the air filter cover ...

... to expose the filter element

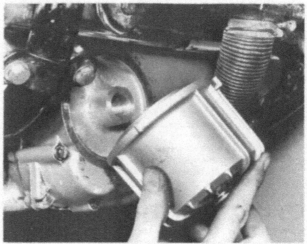

Remove the oil filter housing and element

Replenish engine with oil through filler orifice

check the settings. Rotate the engine through 360° until the 'T - 1' mark is agaiin aligned with the index marks. No 2 piston is now at TDC on the compression stroke. Check, and adjust where necessary, the valve clearances on the following valves.

No 2 inlet	0.1 mm (0.004 in)
No 2 exhaust	0.13 mm (0.005 in)
No 4 exhaust	0.13 mm (0.005 in)
No 3 inlet	0.1 mm (0.004 in)

Check that all the adjuster screw locknuts have been tightened fully. The recommended torque setting is 1.2 - 1.6 kgf m (9 - 12 lbf ft). Refit both cylinder head covers, making sure the rubber seals are correctly positioned, and tighten the retaining bolts to a torque of 0.8 - 1.2 kgf m (6 - 9 lbf ft). Tighten the bolts evenly and in a diagonal sequence.

4   Valve clearance adjustment

Remove the cover from each cylinder head by loosening the retaining bolts evenly and in a diagonal sequence to prevent distortion of the cover. Remove the sparking plugs.

Check the valve clearances on both cylinder heads and adjust, if necessary, as follows. Note that all adjustments must be made with the engine cold.

Rotate the engine until both valves on No 1 cylinder are fully closed and the 'T - 1' mark on the flywheel is aligned with the index marks. With the crankshaft in this position, No 1 piston is at TDC.

Check the following valve clearances by placing a feeler gauge of the required size between the valve stem head and the rocker adjuster screw.

Check valve clearances with feeler gauge

No 1 inlet	0.1 mm (0.004 in)
No 1 exhaust	0.13 mm (0.005 in)
No 3 exhaust	0.13 mm (0.005 in)
No 4 inlet	0.1 mm (0.004 in)

If the gap on any valve is incorrect, loosen the locknut on the adjuster screw and screw the adjuster in or out, as necessary. When adjustment is correct, prevent the screw rotating by using a screwdriver and tighten the locknut. Re-

5   Final drive assembly

Place the machine on the centre stand and remove the oil filler cap from the final drive gear housing. The oil level should be up to the filler orifice. Replenish, if necessary, with a hypoid gear oil conforming to API GL-5 specification. The viscosity of the oil is fairly critical and should be as follows for different ambient temperatures.

Above 5°C (41°F)	SAE 90
Below 5°C (41°F)	SAE 80

Allow the level of the oil to settle before rechecking it. Inspect the O-ring of the filler cap for signs of damage and deterioration and renew it if necessary before fitting and tightening the cap.

Prise the rubber boot off the drive shaft torque tube and check the universal joint for wear. This can be done by holding the gearbox end of the joint and rotating the rear wheel backwards and forwards. There should be no movement between the two 'knuckles' of the joint. If wear is evident, the drive shaft should be renewed. The splined portions of the drive shaft and final drive pinion shaft should also be checked for wear and lubricated, if necessary. This operation and also renewal of the drive shaft, if necessary, due to wear, requires a considerable amount of dismantling. See Chapter 6, Section 19 for the procedures for accomplishing these operations.

Using a grease gun filled with a lithium based multi-purpose grease with a molybdenum disulphide additive, pump grease through the grease nipple located at the final drive housing flange.

6   Fuel lines

Check the fuel lines from the petrol tap to the fuel pump and to the carburettors. If leakage is apparent where the reinforced lines connect at the unions, the hose clips may be tightened. Inspect all lines for chafing or perishing and renew, if suspect.

7   Throttle adjustment

Before checking the throttle twistgrip for the correct amount of free play, rotate the twistgrip to check for smooth operation over its full range with the handlebars positioned at full left lock, full right lock and in the central position. The twistgrip should also automatically close in these positions. If throttle operation is found to be rough or restricted, check the condition and routing of the control cables and lubricate or renew them as necessary.

The amount of free play present in the throttle twistgrip should be 2 - 6 mm (0.08 - 0.20 in). Minor adjustments may be made by turning the cable adjuster at the twistgrip end, loosening the locknut before doing so and tightening it on completion.

If it is not possible to obtain the required amount of free play by turning the upper adjuster, it will be necessary to remove the dummy fuel tank to expose the mid-cable adjuster located just beneath the left-hand electrical component panel. With the upper (fine) adjuster turned fully in, make the necessary adjustment to obtain the free play setting by rotating the mid-cable adjuster. Loosen the locknut before rotating the adjuster and tighten it on completion of adjustment.

8   Choke control adjustment

Operate the choke knob over its full range and check for smooth operation. If any indication of roughness or binding is apparent, check the operating cable for signs of damage and renew the cable if necessary, before lubricating both the cable and choke knob shaft.

With the choke knob pulled fully up, check that the choke valve is fully closed by moving the choke lever located on the rearmost side of No 3 carburettor. If necessary, carry out adjustment by loosening the cable clamp retaining screw and moving the cable outer through the clamp. Once the lever is fully closed, retighten the screw.

With the choke knob pushed fully down, check that the choke valve is fully open by checking for free play in the cable inner where it is exposed between the cable outer and the choke lever. Carry out a final check to determine that the choke knob stays where it is positioned. If this is not the case, remove the rubber cover located beneath the knob to expose the friction adjuster and turn the adjuster the required amount.

Adjust choke cable by loosening clamp retaining screw

9   Carburettor adjustment – synchronisation

In order to maintain optimum engine performance and fuel economy, it is imperative that the carburettors are working in perfect harmony with each other. If the carburettors are not synchronised, not only will one cylinder be doing less work, at any given throttle opening, but it will also in effect have to be carried by the other cylinders. This will reduce performance accordingly.

For synchronisation, it is essential to use a vacuum gauge set consisting of four separate dial gauges, one of which is connected to each carburettor by means of a special adaptor tube. The adaptor pipe screws into the outside lower end of each inlet manifold, the orifice of which is normally blocked off by a cross-head screw plug.

A suitable vacuum gauge set may be purchased from a Honda Service Agent, or from one of the many suppliers who advertise regularly in the motorcycle press. Bear in mind that this equipment is not cheap, and unless the machine is regarded as a long-term purchase, or it is envisaged that similar multi-cylinder motorcycles are likely to follow it, it may be better to allow a Honda dealer to carry out the work. The cost can be reduced considerably if a vacuum gauge set is purchased jointly by a number of owners. As it will be used fairly infrequently this is probably a sound approach.

Place the vacuum gauge set on the machine so that the dials can be easily read. The usual position is between the handlebars. Remove the four blanking plugs from the manifolds and fit the adaptor pieces. Connect the gauges as follows for ease of observation.

No 1 cylinder	Right-hand gauge
No 2 cylinder	Inner right-hand gauge
No 3 cylinder	Inner left-hand gauge
No 4 cylinder	Left-hand gauge

Start the engine allowing it to warm up, and set the speed to 950 ± 100 rpm by turning the throttle stop screw.

Note that if the readings on the gauges fluctuate wildly, it is likely that the gauges require heavier damping. Refer to the gauge manufacturer's instructions on setting up procedures.

Note the reading on each gauge. Select the gauge which shows the reading nearest to midway between the highest and lowest readings of the four gauges and use this reading as a datum upon which to set the other three carburettors.

Pressure between single carburettors in the same pair may be varied by means of the adjustment screw and locknut located between the instruments. Moving this screw in one direction will raise the vacuum on one carburettor and reduce it on the other. Once the single carburettors in the same pair have been set to give as close a reading as possible on their corresponding gauges, the adjustment screw on the rearmost face of No 4 carburettor may be used to regulate the pressures between the two pairs of carburettors.

With all four gauges reading as close to the selected datum as possible, set the engine idle speed to 950 ± 100 rpm and stop the engine. Remove the adaptors, pipes and gauges and refit the manifold blanking plugs. A difference in pressure between all four carburettors of less than 50 mm Hg (2.0 in Hg) should be aimed at if smooth running is to be obtained.

10 Brake system inspection

Carefully inspect both front and rear brake hydraulic lines for signs of damage or deterioration and renew where necessary. The hydraulic lines should be thoroughly cleaned before inspection and renewed if there is the slightest doubt as to their condition. Inspect the unions of both brake systems for signs of leakage, both around the sealing washers and where the union is bonded to the hose. If leakage is apparent, the problem must be dealt with immediately.

Check that all brake system components are properly secured to their respective mounting points and operate both front and rear brake levers to check for full and smooth movement; if necessary, lubricate the lever pivot points. Measure the distance between the top edge of the footrest and the top edge of the rear brake pedal footplate. If the brake pedal height is correct, this distance will be 7 mm (0.27 in). It should be noted that if this distance is not much less than specified, it is likely that the brake will drag because the rider's foot at rest will inadvertently depress the pedal. Adjustment can be carried out by loosening the locknut on the master cylinder pushrod and turning the pushrod to obtain the correct pedal height. On completion of adjustment, re-tighten the locknut and adjust the stop lamp operating switch by referring to the information given in the following Section.

Loosen locknut before turning master cylinder pushrod to obtain correct brake pedal height

11 Stop lamp switch

The rear brake stop lamp switch is adjustable and requires adjustment whenever the brake pedal height is altered. It is a legal requirement that the stop lamp operates whenever the brake pedal is depressed; it should therefore always be correctly adjusted by the following method.

If the stop lamp is late in operating, hold the body still and rotate the combined adjusting mounting nut in an anti-clockwise direction so that the body moves away from the brake pedal shaft. If the switch operates too early or has a tendency to stick on, rotate the nut in a clockwise direction. As a guide, the light should operate after the brake pedal has been depressed by about 2 cm (0.75 in).

A stop lamp switch is also incorporated in the front brake system. The mechanical switch is a push fit in the handlebar lever stock. If the switch malfunctions, repair is impracticable. The switch should be renewed.

12 Steering head bearings

Place the machine on the centre stand so that the front wheel is clear of the ground. This may necessitate placing blocks under the front of the engine. Check the adjustment of the steering head bearings by grasping the forks near the front wheel spindle and pulling and pushing them horizontally in a fore and aft direction. Any movement felt between the steering head lug and the fork yokes indicates that the steering head bearings require adjustment as follows.

Remove the handlebars from the upper fork yoke brackets and allow them to rest on a pad placed on the dummy fuel tank. Slacken the pinch bolt which passes through the rear of the upper fork yoke and the pinch bolt which retains each fork leg to the yoke. Prise the rubber cap from the top of the steering stem and using a C-spanner, remove the slotted nut from the stem, followed by the plain washer. Refer to Section 3 of Chapter 5 and remove the air hose from the top of each fork leg. Detach the instrument assembly from the upper yoke and with the aid of a soft-faced mallet, tap the yoke upwards and off the steering stem so as to expose the locknut and adjuster nut located beneath it. Knock back the tabs of the lockwasher from the slots of the locknut and using a C-spanner, loosen and remove the locknut. Remove and discard the lockwasher, replacing it with a new item.

Adjustment of the steering head bearings requires a socket-type peg spanner so that the slotted adjuster nut can be set to the prescribed torque loading. This requires the use of the special Honda steering stem socket (Part No 07916-3710100) or a home-made equivalent. A piece of tubing can be filed to fit the nut and then welded to a damaged socket to improvise.

Tighten the adjuster nut to a torque of 1.4 - 1.6 kgf m (10 - 12 lbf ft) and turn the steering stem from lock to lock five times to allow the bearings to seat properly. Repeat this sequence twice. Should the nut fail to tighten after turning the steering stem the first or second time, remove the nut and inspect both the stem and nut threads for burrs or contamination by dirt or grit.

On completion of adjustment, grip the front fork lower legs and pivot the steering head from lock to lock. If the bearings feel rough or notchy, then they have become worn or pitted or in extreme cases have fractured. In either event the steering head bearings should be renewed, necessitating complete removal of the forks as described in Chapter 5, Section 5.

Reassembly of the steering head components is a reversal of the dismantling sequence. Refer to the information given in the relevant Sections of Chapter 5 for torque wrench settings, reassembly of the upper yoke components, refitting of the air hose and recharging of the fork legs.

13 Suspension

Carry out a general inspection of the front fork legs, rear suspension units and the swinging arm assembly, check-tightening all nuts and bolts as they are inspected. Check for any damage and signs of leakage and renew or repair each unserviceable component as necessary, referring to the service information given in Chapter 5. Inspect the air hoses for signs of damage and deterioration and ensure that the air pressure in the fork legs and rear suspension units is within the limits listed in Chapter 5.

Check for wear in the swinging arm bearings by placing the machine on the centre stand, with the rear wheel clear of the ground, and pushing and pulling the swinging arm fork ends from side-to-side. The swinging arm is supported on adjustable taper roller bearings, and thus, if the bearings are generally in good condition the play can be taken up. To gain access to the swinging arm so that further investigation and adjustment can be made, remove the rear wheel as described in Chapter 6, Section 11. Temporarily refit the left-hand suspension unit lower bolt so that the weight of the swinging arm is supported. Loosen and remove the three bolts that secure the final drive housing to the tube flange, and lift the housing from position.

Swinging arm adjustment is made by means of the left-hand threaded pivot stub which is secured by a slotted lock ring. To gain access to the stub prise off the plastic inspection cap. Although a soft-metal punch can be used to loosen the lock ring a suitable peg-spanner is required to tighten it after adjustment. Honda provide a special tool No 07908-4690001, or alternatively a tool can be fabricated as described in Chapter 5, Section 9. Fit a suitable Allen key to the left-hand stub and torque load the stub to 8.0 - 12.0 kgf m (58 - 87 lbf ft). By doing this the play should be taken up and the correct pre-load given to the taper bearings. Move the swinging arm through its arc of travel several times and then recheck the torque loading. If the movement of the swinging arm feels notchy, indicating rough bearings, or if play has not been eliminated the swinging arm should be removed for visual inspection of the bearings and cleaning of the adjuster stub threads. It is self-evident that if the adjuster stub is not moving easily in its threads, the correct pre-load on the bearings cannot be applied. Refer to Chapter 5, Section 9 for the full procedure. After completion of adjustment hold the Allen key steady so that the adjuster stub does not turn and tighten the lock ring using the Honda tool or the fabricated item. Refit all displaced components and install the rear wheel.

As a final check of the suspension, test ride the machine, checking that the warning light for the rear suspension air pressure operates correctly before riding away. The warning light should come on for 5 seconds immediately after switching the ignition on; it should then go out. If the light stays on when riding over 10 mph, stop the machine and recheck the pressure in the system. If the light fails to illuminate or comes on but fails to go out, check the pressure warning system by referring to Section 14 of Chapter 7.

Use an accurate gauge when checking air pressure within suspension units

14 Wheels

Inspect both front and rear wheel rims for localised damage in the form of dents or cracks. The existence of even a small crack renders the wheel unfit for further use unless it is found that a permanent repair is possible using arc-welding. This method of repair is highly specialised and therefore the advice of a wheel repair specialist should be sought.

Because tubeless tyres are used, dents may prevent complete sealing between the rim and tyre bead. This may not be immediately obvious until the tyre strikes a severely irregular surface, when the unsupported tyre wall may be deflected away from the rim, causing rapid deflation of the tyre.

Honda recommend that the wheel be renewed if the bead

seating surface of the rim is scratched to a depth of 0.5 mm (0.02 in) or more. Again, if in doubt, seek specialist advice as to whether continued use of the wheel is advisable. Inspect the spoke blades for cracking and security. Check carefully the area immediately around the rivets which pass through the spokes and into the rim.

With the machine placed on the centre stand, check each wheel in turn for rim alignment by placing a pointer close to the rim edge and spinning the wheel.

If the total radial or axial alignment variation is greater than 2.00 mm (0.08 in) the manufacturers recommend that the wheel is renewed. This policy is, however, a counsel of perfection and in practice a larger runout may not affect the handling properties excessively.

Although Honda do not offer any form of wheel rebuilding facility, a number of private engineering firms offer this service. It should be noted however, that Honda do not approve of this course of action.

15 Centre and prop stand

The centre stand pivots on a hollow tube, clamped between two split lugs below the rear engine mounting. The pinch bolts and split-pin should be periodically checked for security and the pivot oiled.

Check also that the return springs have not stretched and that the linkage is unworn; the stand must retract smartly and be firmly held in the retracted position by the springs. If the stand falls whilst the machine is in motion, it may well catch on the road surface causing the rider to be thrown from the machine.

The prop stand pivots on a frame lug located beneath the lower mid-point engine mounting on the left-hand side of the machine. Check the pivot bolt and nut for security and signs of wear and oil the pivot assembly. Check that the return spring has not stretched and that it returns the stand to the retracted position smartly and holds it in place. An accident will almost inevitably result from the prop stand dropping whilst the machine is in motion.

To prevent the machine from being ridden away with the prop stand down, the GL 1100 model incorporates a self-retracting device. Check the rubber 'trip' of this device for wear or damage. No part should be worn below the moulded line on the rubber. Check the operation of the stand as follows. Put the machine on the centre stand, and put the prop stand down. Using a spring balance attached to the extreme end of the centre stand, measure the force required to retract the stand. If this force exceeds 1.5 - 2.5 kg (3.3 - 5.5 lb), check that the stand pivot bolt is not overtight, or in need of lubrication. To renew the rubber 'trip', take off the bolt. Make sure the sleeve is installed in the fixing hole of the new trip. Fit the trip with the arrow facing outwards. The block should be of the type marked 'over 260 lbs only'.

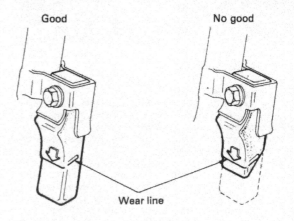
Checking side-stand pad wear

Check propstand rubber 'trip' for wear or damage

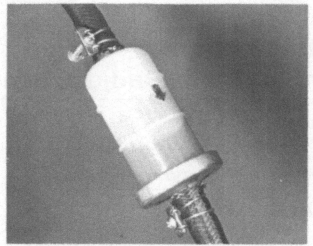

Arrow on filter casing must point in direction of fuel flow

16 Headlamp adjustment

Refer to Sections 17 and 18 of Chapter 7 for details of the method of headlamp adjustment on the Standard and Interstate models respectively. Remember that not only is it a legal requirement to have a correctly adjusted headlamp but that a headlamp set to dazzle oncoming motorists represents not only a danger to that motorist but also a danger to any person or vehicle that he might be incapable of seeing.

---

2 yearly, or every 24 000 miles (38 400 km)

---

1 Fuel filter renewal

The fuel filter is a sealed unit and therefore cleaning is impracticable; a new unit must be fitted. The filter is secured by a clamp bracket, which is retained by one bolt to the lower seam of the fuel tank. Remove the filter by separating the clamp and pulling the fuel feed hose off each end of the filter, after loosening the screw clips.

Note that the new filter unit must be fitted with the arrow on its casing pointing in the direction of fuel flow, ie towards the fuel pump. On completion of fitting the unit, start the engine and check for fuel leaks at the disturbed fuel hose connections.

2 Brake fluid

Honda recommend that the brake fluid in both front and rear brake systems be renewed at this service interval. It is practical to combine this service operation with a complete internal and external examination of the brake operating components, overhauling and renewing where necessary. Refer to Chapter 6 for details of these operations.

3 Coolant

It is recommended that, at this service interval, the engine cooling system is completely drained of coolant, flushed out with a suitable flushing agent and refilled with new coolant. Full details of draining, flushing and refilling the system, as well as the correct bleeding procedure, may be found in Sections 2, 3 and 4 of Chapter 2.

4 Final drive lubrication

Position the machine on its centre stand and place a suitable container below the final drive gear housing. Remove the filler cap followed by the drain plug and allow all the oil contained within the housing to drain into the container. Any oil left within the housing may be expelled by turning the rear wheel by hand.

Inspect the condition of both the filler cap O-ring and the drain plug sealing washer, renewing either one if necessary.

Refit and tighten the drain bolt to 1.0 - 1.4 kgf m (7 - 10 lbf ft). Refill the housing with 140 - 160 cc (5.07/5.28 US/Imp fl oz) of hypoid gear oil conforming to API GL-5 specification. In the UK a good quality EP hypoid gear oil should be used. If the normal ambient temperature in which the machine operates is less than 5°C (41°F), use an SAE 80 oil. If the normal temperature is greater than this, use SAE 90 oil. Take great care when refilling not to allow foreign matter into the housing and avoid spilling the lubricant on the rear tyre or brake components.

Before refitting the filler cap, allow three minutes for the oil to find its way around the gear teeth and bearings. Recheck that the level is up to the filler orifice and refit and tighten the filler cap to a torque of 1.0 - 1.4 kgf m (7 - 10 lbf ft).

Additional service items

From time to time the exposed ends of all the control cables should be lubricated with light oil to inhibit corrosion and to help prevent the ingress of water.

At greater intervals the control cables should be lubricated more thoroughly using a hydraulic cable oiler when possible. This method of lubrication ensures that lubricant finds its way along the whole length of cable. If a hydraulic oiler is not available use the alternative method shown in the accompanying illustration. This method has the disadvantage the cables have to be removed from the machine to allow the oil to drip through. Nylon lined cables, which may have been fitted as replacements, should not be lubricated; in some cases the oil will cause the inner casing to swell leading to cable seizure.

Oiling a control cable

Quick glance
maintenance adjustments and capacities

Engine/gearbox oil capacity
    After draining .... 3.2 litres (3.4/2.8 US/Imp quarts)
    After dismantling .... 4.0 litres (4.2/3.5 US/Imp quarts)

Final drive bevel housing oil
    Capacity .... 140 – 160 cc (5.07/5.28 US/Imp fl oz)

Cooling system
    Capacity .... 3.4 litres (3.6/3.0 US/Imp quarts)

Sparking plug gap .... 0.6 – 0.7 mm (0.024 - 0.028 in)

Valve clearances (cold)
    Inlet .... 0.10 mm (0.004 in)
    Exhaust .... 0.13 mm (0.005 in)

Tyre pressures
    Front and rear* .... 32 psi (2.25 kg/cm^2)
Above 200 lb (90 kg) load increase rear tyre pressure to 40 psi (2.80 kg/cm^2)

Front fork oil capacity (per leg)
    UK model .... 220 cc (7.74 Imp fl oz)
    US models .... 240 cc (8.11 US fl oz)

Suspension air pressures
    Front .... 14 – 21 psi (1.0 – 1.5 kg/cm^2)
    Rear:
        UK and 1980 (US) models .... 28 – 42 psi (2.0 – 3.0 kg/cm^2)
        1981 (US) models .... 28 – 57 psi (2.0 – 4.0 kg/cm^2)

Recommended lubricants

Component	Lubricant
Engine/gearbox	SAE 10W/40 engine oil
Final drive unit	Hypoid gear oil, SAE 80 or SAE 90
Front forks and rear suspension units	Automatic transmission fluid (ATF)
Wheel bearings	High melting point grease
Steering head bearings	High melting point grease
Swinging arm bearings	High melting point grease
Main and prop stand pivots	Multi-purpose grease
Brake pedal pivot	Multi-purpose grease
Final drive shaft	Lithium based multi-purpose grease with molybdenum disulphide additive
Hydraulic brakes	DOT 3 (US) or SAE J1703 (UK) hydraulic brake fluid

Conversion factors

Length (distance)

Inches (in)	X	25.4	= Millimetres (mm)	X	0.0394	= Inches (in)
Feet (ft)	X	0.305	= Metres (m)	X	3.281	= Feet (ft)
Miles	X	1.609	= Kilometres (km)	X	0.621	= Miles

Volume (capacity)

Cubic inches (cu in; in³)	X	16.387	= Cubic centimetres (cc; cm³)	X	0.061	= Cubic inches (cu in; in³)
Imperial pints (Imp pt)	X	0.568	= Litres (l)	X	1.76	= Imperial pints (Imp pt)
Imperial quarts (Imp qt)	X	1.137	= Litres (l)	X	0.88	= Imperial quarts (Imp qt)
Imperial quarts (Imp qt)	X	1.201	= US quarts (US qt)	X	0.833	= Imperial quarts (Imp qt)
US quarts (US qt)	X	0.946	= Litres (l)	X	1.057	= US quarts (US qt)
Imperial gallons (Imp gal)	X	4.546	= Litres (l)	X	0.22	= Imperial gallons (Imp gal)
Imperial gallons (Imp gal)	X	1.201	= US gallons (US gal)	X	0.833	= Imperial gallons (Imp gal)
US gallons (US gal)	X	3.785	= Litres (l)	X	0.264	= US gallons (US gal)

Mass (weight)

Ounces (oz)	X	28.35	= Grams (g)	X	0.035	= Ounces (oz)
Pounds (lb)	X	0.454	= Kilograms (kg)	X	2.205	= Pounds (lb)

Force

Ounces-force (ozf; oz)	X	0.278	= Newtons (N)	X	3.6	= Ounces-force (ozf; oz)
Pounds-force (lbf; lb)	X	4.448	= Newtons (N)	X	0.225	= Pounds-force (lbf; lb)
Newtons (N)	X	0.1	= Kilograms-force (kgf; kg)	X	9.81	= Newtons (N)

Pressure

Pounds-force per square inch (psi; lbf/in²; lb/in²)	X	0.070	= Kilograms-force per square centimetre (kgf/cm²; kg/cm²)	X	14.223	= Pounds-force per square inch (psi; lbf/in²; lb/in²)
Pounds-force per square inch (psi; lbf/in²; lb/in²)	X	0.068	= Atmospheres (atm)	X	14.696	= Pounds-force per square inch (psi; lbf/in²; lb/in²)
Pounds-force per square inch (psi; lbf/in²; lb/in²)	X	0.069	= Bars	X	14.5	= Pounds-force per square inch (psi; lbf/in²; lb/in²)
Pounds-force per square inch (psi; lbf/in²; lb/in²)	X	6.895	= Kilopascals (kPa)	X	0.145	= Pounds-force per square inch (psi; lbf/in²; lb/in²)
Kilopascals (kPa)	X	0.01	= Kilograms-force per square centimetre (kgf/cm²; kg/cm²)	X	98.1	= Kilopascals (kPa)
Millibar (mbar)	X	100	= Pascals (Pa)	X	0.01	= Millibar (mbar)
Millibar (mbar)	X	0.0145	= Pounds-force per square inch (psi; lbf/in²; lb/in²)	X	68.947	= Millibar (mbar)
Millibar (mbar)	X	0.75	= Millimetres of mercury (mmHg)	X	1.333	= Millibar (mbar)
Millibar (mbar)	X	0.401	= Inches of water (inH₂O)	X	2.491	= Millibar (mbar)
Millimetres of mercury (mmHg)	X	0.535	= Inches of water (inH₂O)	X	1.868	= Millimetres of mercury (mmHg)
Inches of water (inH₂O)	X	0.036	= Pounds-force per square inch (psi; lbf/in²; lb/in²)	X	27.68	= Inches of water (inH₂O)

Torque (moment of force)

Pounds-force inches (lbf in; lb in)	X	1.152	= Kilograms-force centimetre (kgf cm; kg cm)	X	0.868	= Pounds-force inches (lbf in; lb in)
Pounds-force inches (lbf in; lb in)	X	0.113	= Newton metres (Nm)	X	8.85	= Pounds-force inches (lbf in; lb in)
Pounds-force inches (lbf in; lb in)	X	0.083	= Pounds-force feet (lbf ft; lb ft)	X	12	= Pounds-force inches (lbf in; lb in)
Pounds-force feet (lbf ft; lb ft)	X	0.138	= Kilograms-force metres (kgf m; kg m)	X	7.233	= Pounds-force feet (lbf ft; lb ft)
Pounds-force feet (lbf ft; lb ft)	X	1.356	= Newton metres (Nm)	X	0.738	= Pounds-force feet (lbf ft; lb ft)
Newton metres (Nm)	X	0.102	= Kilograms-force metres (kgf m; kg m)	X	9.804	= Newton metres (Nm)

Power

Horsepower (hp)	X	745.7	= Watts (W)	X	0.0013	= Horsepower (hp)

Velocity (speed)

Miles per hour (miles/hr; mph)	X	1.609	= Kilometres per hour (km/hr; kph)	X	0.621	= Miles per hour (miles/hr; mph)

Fuel consumption*

Miles per gallon, Imperial (mpg)	X	0.354	= Kilometres per litre (km/l)	X	2.825	= Miles per gallon, Imperial (mpg)
Miles per gallon, US (mpg)	X	0.425	= Kilometres per litre (km/l)	X	2.352	= Miles per gallon, US (mpg)

Temperature

Degrees Fahrenheit = (°C x 1.8) + 32   Degrees Celsius (Degrees Centigrade; °C) = (°F - 32) x 0.56

*It is common practice to convert from miles per gallon (mpg) to litres/100 kilometres (l/100km),
where mpg (Imperial) x l/100 km = 282 and mpg (US) x l/100 km = 235

Working conditions and tools

When a major overhaul is contemplated, it is important that a clean, well-lit working space is available, equipped with a workbench and vice, and with space for laying out or storing the dismantled assemblies in an orderly manner where they are unlikely to be disturbed. The use of a good workshop will give the satisfaction of work done in comfort and without haste, where there is little chance of the machine being dismantled and reassembled in anything other than clean surroundings. Unfortunately, these ideal working conditions are not always practicable and under these latter circumstances when improvisation is called for, extra care and time will be needed.

The other essential requirement is a comprehensive set of good quality tools. Quality is of prime importance since cheap tools will prove expensive in the long run if they slip or break when in use, causing personal injury or expensive damage to the component being worked on. A good quality tool will last a long time, and more than justify the cost.

For practically all tools, a tool factor is the best source since he will have a very comprehensive range compared with the average garage or accessory shop. Having said that, accessory shops often offer excellent quality tools at discount prices, so it pays to shop around. There are plenty of tools around at reasonable prices, but always aim to purchase items which meet the relevant national safety standards. If in doubt, seek the advice of the shop proprietor or manager before making a purchase.

The basis of any tool kit is a set of open-ended spanners, which can be used on almost any part of the machine to which there is reasonable access. A set of ring spanners makes a useful addition, since they can be used on nuts that are very tight or where access is restricted. Where the cost has to be kept within reasonable bounds, a compromise can be effected with a set of combination spanners – open-ended at one end and having a ring of the same size on the other end. Socket spanners may also be considered a good investment, a basic $^3/_8$ in or $^1/_2$ in drive kit comprising a ratchet handle and a small number of socket heads, if money is limited. Additional sockets can be purchased, as and when they are required. Provided they are slim in profile, sockets will reach nuts or bolts that are deeply recessed. When purchasing spanners of any kind, make sure the correct size standard is purchased. Almost all machines manufactured outside the UK and the USA have metric nuts and bolts, whilst those produced in Britain have BSF or BSW sizes. The standard used in USA is AF, which is also found on some of the later British machines. Others tools that should be included in the kit are a range of crosshead screwdrivers, a pair of pliers and a hammer.

When considering the purchase of tools, it should be remembered that by carrying out the work oneself, a large proportion of the normal repair cost, made up by labour charges, will be saved. The economy made on even a minor overhaul will go a long way towards the improvement of a toolkit.

In addition to the basic tool kit, certain additional tools can prove invaluable when they are close to hand, to help speed up a multitude of repetitive jobs. For example, an impact screwdriver will ease the removal of screws that have been tightened by a similar tool, during assembly, without a risk of damaging the screw heads. And, of course, it can be used again to retighten the screws, to ensure an oil or airtight seal results. Circlip pliers have their uses too, since gear pinions, shafts and similar components are frequently retained by circlips that are not too easily displaced by a screwdriver. There are two types of circlip pliers, one for internal and one for external circlips. They may also have straight or right-angled jaws.

One of the most useful of all tools is the torque wrench, a form of spanner that can be adjusted to slip when a measured amount of force is applied to any bolt or nut. Torque wrench settings are given in almost every modern workshop or service manual, where the extent to which a complex component, such as a cylinder head, can be tightened without fear of distortion or leakage. The tightening of bearing caps is yet another example. Overtightening will stretch or even break bolts, necessitating extra work to extract the broken portions.

As may be expected, the more sophisticated the machine, the greater is the number of tools likely to be required if it is to be kept in first class condition by the home mechanic. Unfortunately there are certain jobs which cannot be accomplished successfully without the correct equipment and although there is invariably a specialist who will undertake the work for a fee, the home mechanic will have to dig more deeply in his pocket for the purchase of similar equipment if he does not wish to employ the services of others. Here a word of caution is necessary, since some of these jobs are best left to the expert. Although an electrical multimeter of the AVO type will prove helpful in tracing electrical faults, in inexperienced hands it may irrevocably damage some of the electrical components if a test current is passed through them in the wrong direction. This can apply to the synchronisation of twin or multiple carburettors too, where a certain amount of expertise is needed when setting them up with vacuum gauges. These are, however, exceptions. Some instruments, such as a strobe lamp, are virtually essential when checking the timing of a machine powered by CDI ignition system. In short, do not purchase any of these special items unless you have the experience to use them correctly.

Although this manual shows how components can be removed and replaced without the use of special service tools (unless absolutely essential), it is worthwhile giving consideration to the purchase of the more commonly used tools if the machine is regarded as a long term purchase Whilst the alternative methods suggested will remove and replace parts without risk of damage, the use of the special tools recommended and sold by the manufacturer will invariably save time.

Chapter 1 Engine, clutch and gearbox

Contents

Specifications

Engine

Type	Horizontally-opposed, water cooled, OHC, 4 cylinder, 4 stroke
Bore	75 mm (2.953 in)
Stroke	61.4 mm (2.417 in)
Compression ratio	9.2 : 1
Capacity	1085 cc (66.2 cu in)
Compression pressure	12 kg/sq cm (171 psi)
Engine rotation	Clockwise (viewed from front of engine)

Camshaft

End bearing ID	27.0 – 27.021 mm (1.0630 – 1.0638 in)
Service limit	27.050 mm (1.0650 in)
End journal OD	26.954 – 26.970 mm (1.0612 – 1.0618 in)
Service limit	26.910 mm (1.0594 in)
Centre bearing ID	25.0 – 25.021 mm (0.9843 – 0.9851 in)
Service limit	25.050 mm (0.9862 in)
Centre journal OD	24.934 – 24.950 mm (0.9817 – 0.9823 in)
Service limit	24.910 mm (0.9807 in)
End bearing clearance	0.030 – 0.067 mm (0.0011 – 0.0026 in)
Service limit	0.140 mm (0.0055 in)
Centre bearing clearance	0.050 – 0.087 mm (0.0020 – 0.0034 in)
Service limit	0.140 mm (0.0055 in)
Height:	
Inlet	37.0 mm (1.46 in)
Service limit	36.8 mm (1.45 in)
Exhaust	36.8 mm (1.45 in)
Service limit	36.6 mm (1.44 in)

Valves and springs

Operation	Toothed belt driven single overhead camshaft
Timing (with camshaft at 1 mm lift):	
Inlet opens	5° BTDC
Inlet closes	43° ABDC
Exhaust opens	45° BBDC
Exhaust closes	5° ATDC
Clearances (Cold):	
Inlet	0.1 mm (0.004 in)
Exhaust	0.13 mm (0.005 in)
Valve to guide clearances:	
Inlet	0.01 – 0.04 mm (0.0004 – 0.0016 in)
Service limit	0.08 mm (0.0032 in)
Exhaust	0.05 – 0.07 mm (0.0020 – 0.0028 in)
Service limit	0.10 mm (0.0039 in)
Guide ID:	
Inlet and exhaust	6.60 – 6.62 mm (0.2598 – 0.2606 in)
Service limit	6.64 mm (0.2614 in)
Valve seat width	1.40 mm (0.0551 in)
Spring free length:	
Inner	40.2 mm (1.5827 in)
Service limit	39.0 mm (1.5354 in)
Outer	43.75 mm (1.7225 in)
Service limit	42.5 mm (1.6732 in)

Rockers

Arm ID	14.0 – 14.018 mm (0.5512 – 0.5519 in)
Service limit	14.05 mm (0.5532 in)
Shaft OD	13.966 – 13.984 mm (0.5498 – 0.5505 in)
Service limit	13.94 mm (0.5488 in)

Cylinder head

Maximum warpage	0.1 mm (0.0039 in)

Cylinders

Bore ID	75.0 – 75.015 mm (2.9528 – 2.9533 in)
Service limit	75.10 mm 2.9567 in)
Bore ovality (max)	0.15 mm (0.0059 in)
Bore taper	0.007 – 0.012 mm (0.0003 – 0.0004 in)
Service limit	0.05 mm (0.002 in)
Cylinder to piston clearance	0.020 – 0.065 mm (0.0008 – 0.0026 in)
Service limit	0.25 mm (0.0098 in)

Pistons and rings
Skirt OD .... 74.950 – 74.980 mm (2.9508 – 2.9520 in)
Service limit .... 74.850 mm (2.9468 in)
Ring end gap:
   Top and second .... 0.1 – 0.3 mm (0.004 – 0.012 in)
   Service limit .... 0.6 mm (0.02 in)
   Oil ring side rail .... 0.2 – 0.9 mm (0.008 – 0.035 in)
   Service limit .... 1.10 mm (0.04 in)
Ring groove width:
   Top and second .... 1.205 – 1.220 mm (0.0475 – 0.0480 in)
   Service limit .... 1.30 mm (0.0512 in)
   Oil ring .... 2.505 – 2.520 mm (0.0986 – 0.0992 in)
   Service limit .... 2.60 mm (0.1024 in)
Ring to groove clearance:
   Top and second .... 0.015 – 0.045 mm (0.0006 – 0.0018 in)
   Service limit .... 0.120 mm (0.0047 in)

Big-end bearings
Radial clearance .... 0.020 – 0.044 mm (0.0008 – 0.0017 in)
Service limit .... 0.080 mm (0.0031 in)
Axial clearance .... 0.15 – 0.30 mm (0.0059 – 0.0118 in)
Service limit .... 0.40 mm (0.0158 in)

Crankshaft
Maximum runout .... 0.05 mm (0.0019 in)
Journal ovality .... 0.005 mm (0.0002 in)
Service limit .... 0.008 mm (0.0003 in)
Journal taper .... 0.002 mm (0.00008 in)
Service limit .... 0.004 mm (0.0002 in)
Journal radial clearance .... 0.02 – 0.044 mm (0.0008 – 0.0017 in)
Service limit .... 0.08 mm (0.0032 in)
Main bearing radial clearance .... 0.020 – 0.044 mm (0.0008 – 0.0017 in)
Service limit .... 0.080 mm (0.0031 in)

Clutch
Type .... Wet, multi-plate
No of plates:
   Friction 8
   Plain 6
Friction plate thickness .... 3.42 – 3.58 mm (0.1347 – 0.1410 in)
Service limit .... 3.20 mm (0.1260 in)
Maximum plate warpage .... 0.15 mm (0.006 in)
Springs:
   No of 6
   Free length .... 35.5 mm (1.3976 in)
   Service limit .... 34.2 mm (1.3386 in)

Gearbox
Type .... 5-speed, constant mesh
Primary reduction ratio .... 1.708 : 1 (41/24)
Secondary reduction ratio .... 0.973 : 1 (36/37)
Final reduction ratio .... 3.091 : 1 (34/11)
Gear ratios:
   1st .... 2.500 : 1 (40/16)
   2nd .... 1.667 : 1 (40/24)
   3rd .... 1.286 : 1 (36/28)
   4th .... 1.065 : 1 (33/31)
   5th .... 0.909 : 1 (30/33)
Primary transmission .... Hy Vo chain
Selector fork ID .... 13.0 – 13.027 mm (0.5118 – 0.5129 in)
Service limit .... 13.050 mm (0.5138 in)
Selector fork shaft OD .... 12.966 – 12.984 mm (0.5105 – 0.5112 in)
Service limit .... 12.90 mm (0.5079 in)
Fork finger thickness .... 6.4 – 6.5 mm (0.2520 – 0.2559 in)
Service limit .... 6.1 mm (0.2402 in)
Gearchange drum OD:
   Inside .... 11.966 – 11.984 mm (0.4711 – 0.4718 in)
   Service limit .... 11.95 mm (0.4705 in)
   Outside .... 35.959 – 35.980 mm (1.41547 – 1.4166 in)
   Service limit .... 35.92 mm (1.4142 in)
Gearchange drum groove width .... 13.0 – 13.018 mm (0.5118 – 0.5125 in)
Service limit .... 13.04 mm (0.5134 in)
Final gear shaft damper spring free length .... 110.9 mm (4.3661 in)
Service limit .... 100.0 mm (3.9370 in)

Torque wrench settings

	kgf m	lbf ft
Cylinder head bolts (10 mm)	5.3 – 5.7	38 – 41
Cylinder head cover special bolts (6 mm)	0.8 – 1.2	6 – 9
Camshaft holder bolts (8 mm)	1.8 – 2.2	13 – 16
Timing belt cover special bolts (6 mm)	0.8 – 1.2	6 – 9
Timing belt driven pulley bolt (8 mm)	2.5 – 2.9	8 – 21
Timing belt drive pulley bolt (12 mm)	7.0 – 8.0	51 – 58
Timing belt tensioner rubber bolt (8 mm)	2.4 – 2.8	17 – 20
Connecting rod big-end cap nuts (8 mm)	3.0 – 3.4	22 – 25
Centre main bearing cap bolts (12 mm)	6.7 – 7.3	49 – 53
End main bearing cap bolts (10 mm)	4.8 – 5.2	35 – 38
Crankcase bolts (6 mm)	1.0 – 1.4	7 – 10
Crankcase bolts (8 mm)	2.4 – 2.8	17 – 20
Crankcase bolts (10 mm)	3.3 – 3.7	24 – 27
Crankcase sealing bolts (18 mm)	3.5 – 5.5	25 – 40
Valve adjusting screw locknuts	1.2 – 1.6	9 – 12
Alternator rotor centre bolt (12 mm)	8.0 – 9.0	58 – 65
Alternator bearing holder bolts (6 mm)	1.0 – 1.4	7 – 10
Clutch centre locknut (20 mm)	5.5 – 6.5	40 – 47
Clutch spring bolts (6 mm)	0.8 – 1.2	6 – 9
Gearchange arm set bolt (8 mm)	2.3 – 2.7	17 – 20
Gearchange drum stopper arm bolt (6 mm)	1.0 – 1.4	7 – 10
Starting clutch outer screw (8 mm)	2.3 – 2.7	17 – 20
Timing mark cap	1.2 – 1.6	9 – 12
Alternator cover cap	1.0 – 1.4	7 – 10
Engine mounting bolts (12 mm)	5.5 – 6.5	40 – 47
Engine mounting bolts (10 mm)	3.0 – 4.0	22 – 29
Engine mounting bolts (8 mm)	1.8 – 2.5	13 – 18
Gearchange pedal pinch bolt (6 mm)	0.8 – 1.2	6 – 9

1 General description

The engine fitted to the Honda GL1100 is of unusual design, incorporating many features more usual to motor car engine design practice. The engine has four cylinders arranged in two horizontally opposed banks, lying either side of the machine centre line. The aluminium crankcase separates about a vertical plane and houses steel cylinder liners, there being no cylinder barrels as such. The crankshaft runs on three shell type main bearings located by caps bolted to the right-hand crankcase half. The big-end bearings are of similar type. When the crankcases are separated, the crankshaft and gearbox components which are mounted below, remain in the right-hand casing. The gearbox is placed below the crankshaft, to reduce the overall length of the engine unit; primary drive being transmitted by a Hy-vo chain at the rear of the engine. Each cylinder bank shares a common cylinder head, fitted with two offset valves per combustion chamber, and operated by a single camshaft. The camshafts are driven by two separate toothed belts from two pulleys mounted in tandem on the extreme front end of the crankshaft. Belt tension is maintained by manually adjustable jockey pulleys, which are tensioned automatically by springs.

Lubrication is of the high pressure, wet sump type, where all the lubricant is contained within the crankcase. Two pumps are fitted, at either end of a single shaft, driven by a roller chain from a sprocket at the rear of the clutch outer drum. The main oil pump is fitted at the front of the engine and supplies oil to the bearing surfaces of the engine via a car type filter mounted on a detachable housing at the extreme front of the engine. Oil returns to the sump by gravity, except for that trapped in the clutch housing, which is returned by the second pump (clutch scavenge pump).

Power from the engine is transmitted to a multiplate clutch and then via a drive shaft to a crown wheel and pinion contained within an aluminium housing at the rear wheel.

In common with many motor car engines, the GL 1100 engine is water cooled. The coolant, which comprises a 50/50 mixture of water and anti-freeze, is circulated around the engine by means of a water pump driven off the forward end of the oil pump drive shaft. The coolant is then passed through a radiator, mounted forward of the engine on the duplex frame down tubes, where it is cooled and returned to the engine. Incorporated in the system is a thermostat, which helps reduce the warming-up period of the engine and regulates the coolant temperature. An electrically driven fan is also fitted to the rear of the radiator, which cuts in automatically when a preset temperature is reached.

2 Preparation for engine/gearbox removal: Interstate model

In order to give full and unobstructed access to the engine/gearbox unit, it is necessary to remove the fairing panels and mounting bracket assembly fitted to the Interstate model by referring to the service information given in Section 19 of Chapter 5.

3 Operations with the engine/gearbox in the frame

Owing to the unusual engine design, access to most main components is obscured when the engine is in the frame. The following components can be removed with the engine still in situ:

(a) Cylinder heads and valve gear
(b) Timing belts and pulleys
(c) Carburettor assembly
(d) Fuel pump
(e) Starter motor
(f) Clutch lifting mechanism and plates
(g) Main oil pump

4   Operations with engine/gearbox removed from the frame

The engine must be removed for access to and removal of all remaining components, including the following:

(a)   Clutch outer drum
(b)   Alternator
(c)   CDI pulser generator
(d)   Gearchange mechanism
(e)   Gearbox components
(f)   Crankshaft assembly
(g)   Pistons and connecting rods
(h)   Clutch oil scavenge pump

5   Method of engine/gearbox removal

The engine and gearbox are of unit construction (ie housed within the same casing) and as such it is necessary to remove the unit complete in order to gain access to either sub-assembly. Separation of the crankcase is achieved after the engine has been removed from the frame and refitting cannot take place until the engine/gear unit is assembled completely.

6   Engine/gearbox unit: removal from the frame

1   Place the machine firmly on its centre stand so that it is standing securely and there is no likelihood that it may fall over. This is extremely important as owing to the weight of the complete machine and the engine, any instability during dismantling will probably be uncontrollable. If possible, place the machine on a raised platform. This will improve accessibility and ease engine removal. Again, owing to the weight of the machine, ensure that the platform is sufficiently strong and well supported.
2   Remove the two sidepanels by pulling them out of their frame locating grommets (three to each panel). Remove the seat by unscrewing the Allen bolt situated each side at the base of the pillion seat, moving the seat backwards to detach it from the forward frame mounting and lifting the front of the seat clear of the rear of the dummy fuel tank so that it can be pulled forwards and away from the machine. The dummy fuel tank may now be detached from its frame mounting points by unscrewing the four securing bolts, two at the front of the tank and two at the rear. Open the flaps on the tank top and lift out the tool tray. The dummy tank may now be carefully lifted up, bearing it slightly to the right so that it clears the components located within it. Once clear of the frame, the dummy tank, sidepanels and seat should all be carefully stored, away from any potential hazards such as dropped tools, gritty surfaces, etc.
3   Place a container of more than 3.4 litres (about 1 gallon) capacity below the front of the engine, remove the radiator pressure cap and remove the coolant drain plug from the lower edge of the water pump outer casing. Allow the coolant to drain. If the coolant is to be reused, ensure that the container is absolutely clean. The container should ideally be of a plastic material as this will not contaminate the coolant. Place a second container below the engine oil drain plug and remove the plug, allowing all the oil to drain. The plug is situated below the oil filter housing. The container used should be of more than 4.0 litres (about 1.25 gallons) capacity.
4   Disconnect the leads from the battery terminals; positive (+) lead followed by the negative (–) lead. If the machine is to be inoperative for an extended length of time, the battery should be removed by unscrewing the retaining bracket bolt and swinging the bracket down to clear the battery. Pull the vent pipe from the battery union. The battery may now be lifted clear of the machine and placed on a clean work surface so that it can be serviced and charged when convenient.

5   Loosen the lower hose clip on the radiator upper hose and pull the hose off the thermostat housing union. The radiator lower hose is retained in a similar manner. However, owing to its short length, removal is made easier if the manifold is removed from the water pump outer casing. The manifold is retained by two bolts. Disconnect the pipe connecting the expansion tank to the radiator filler point by removing the wire retaining clip and pulling the pipe end off the radiator stub. Disconnect the radiator fan electrical leads by unclipping the block connector. The radiator may now be removed from the machine by removing the two lower mounting bolts, followed by the two upper mounting dome nuts and washers, noting the electrical loom guard bracket located beneath the left-hand nut. The radiator must be supported when removing these mounting nuts and bolts to prevent it falling forward.Turn the forks onto full left-hand lock and carefully ease the radiator forward and sideways away from the machine, taking care not to damage the radiator fins. Store the radiator with the dummy fuel tank, etc.
6   Complete the draining of the engine/gearbox oil by moving the container from beneath the oil drain plug to beneath the oil filter housing. Remove the filter housing centre bolt and pull the housing, complete with filter away from the engine casing. This will allow any oil contained within the housing to drain into the container.
7   The machine featured in this Manual was fitted with crash bars to protect the cylinder heads and covers in the event of the machine being 'dropped'. Removal of these bars was effected by first removing the right-hand U-brackets, nuts and bolts, followed by the single lower locknut and plain washer. This allowed the right-hand section of the crash bar assembly to be pulled away from the left-hand section. Care should be taken to retain the spacer located beneath the lower crash bar mounting. Removal of the left-hand section was by a similar procedure, the mid-point mounting being retained in position by a single dome nut and washer instead of a U-bracket.
8   Removal of the exhaust system may be more easily achieved with the aid of an assistant. Remove the two flange nuts from each of the four exhaust pipe to cylinder head connections. Remove the two bolts from the balancer pipe clamp and check that the clamp is free. Loosen the nut and bolt retaining each silencer to each pillion footrest mounting bracket and remove each nut. With a person each side of the machine supporting the exhaust assembly, withdraw the mounting bolt from each pillion footrest bracket and carefully lower the complete assembly away from its mounting points. Once clear, the balancer pipe may be separated at its connection and the two separate exhaust pipes lifted clear of the machine. In no circumstances should the exhaust assembly to allowed to hang unsupported from its cylinder head mounting points because the weight of the system will place an unacceptable strain on the cylinder head studs.
9   Ensure all electrical leads between the main wiring harness and engine components are disconnected, these include:
The CDI pulser generator block connector
The alternator block connector
The engine main harness block connector
The starter motor cable at the motor terminal
The water temperature gauge sensor switch socket connector
The oil pressure warning switch screw connection
Detach the HT leads from their retaining clips on the carburettor assembly, pull the suppressor caps off the sparking plugs and lift the leads clear of the engine. It should be noted that the lead to the neutral indicator switch cannot be disconnected until the engine is moved clear of the frame.
10 Turn the fuel tap to the 'Off' position and loosen the hose clip which secures the fuel feed at the fuel pump. The pipe will pull from position. Disconnect the tachometer drive cable from the gearbox in the fuel pump mounting casing by removing the inner flange retaining bolt. Refit the bolt after pulling the cable from position.
11 Disconnect the clutch cable from the operating lever at the

engine and by first loosening the cable adjuster nuts, thus allowing the adjuster to pass fully down through its guide, and then lifting the lever to allow the cable nipple to be disconnected from the lower end. Pull the rubber sheath down to expose the inner cable, pass the cable sideways through the gap in the adjuster guide and secure the freed cable clear of the engine.

12 Remove the air filter cover by unscrewing the single retaining wingnut and lifting the cover clear of the filter housing, along with its seal. The filter element and seal may now be withdrawn from the housing. Remove the element retainer by unscrewing its two securing bolts and pulling it out of the filter housing. Finally, unclip the breather pipes from the housing stubs and lift the housing clear of the carburettor manifold inlet, noting the sealing ring at the base of the housing.

13 Disconnect the choke cable from its clamp located just to the rear of the rear right-hand carburettor; the outer cable may be freed simply by unscrewing the crosshead retaining screw. Free the cable nipple from the operating lever.

14 Release the throttle cables by loosening the cable adjuster nuts at the engine end to allow the adjusters to pass fully down through their guides. The cable nipples may now be released from the operating linkage by opening or closing the throttle twistgrip, so that either one of the cables is slack, and using a pair of long-nosed pliers to release the nipple of the slackened cable from the operating pulley. To free the cable ends from their adjuster guides, fully undo the adjuster nuts and pull the adjusters up through the guides so that the cable inners may be moved sideways to clear the guide slots. Secure the freed cables clear of the engine. Access to these cable ends was difficult and some patience was needed before the job of disconnecting them was finally completed. This operation can be accomplished with the engine partially removed from the frame in which case care must be taken not to place any severe strain on the cables.

15 Remove the rear brake stop lamp operating switch spring so that unobstructed access may be gained to the final drive shaft rubber gaiter. Release the gaiter retaining spring and prise the gaiter off the engine casing retaining lip. With the gaiter pulled fully back and using a pair of externally opening 90° cranked jaw circlip pliers, remove the circlip which retains the forward 'knuckle' of the drive shaft universal joint on the engine shaft. This operation requires care, as the circlip is partially obscured. After removal of the circlip, prise the complete joint backwards until it leaves the splines of the engine shaft.

16 Because of the great weight of the engine unit and the small amount of room available for manoeuvering, a hydraulic trolley jack should be used to support the engine as it is removed from the frame. Although it may be possible in practice to lift the engine clear with the aid of two or three assistants, the likelihood of damage to the engine is very great and should be avoided at all costs. The complete engine removal operation is as follows.

17 Place the trolley jack at right angles to the engine centre line, on the left-hand side of the machine. If the machine is supported on a raised platform, an extension to the platform should be arranged upon which the jack can stand and be pulled outwards, bearing the weight of the engine. Place a suitable plank of wood below the engine sump and position the jack so that the engine is slightly supported. Try and position the jack below the engine so that the engine will be balanced on removal. This is largely a matter of guesswork.

18 Remove the left-hand footrest by unscrewing the retaining dome nut; remove also the smaller dome nut to the rear of the footrest. Locate the engine casing to frame attachment bolt, situated directly above the prop stand pivot, and remove it. Move to the forward end of the removable cradle frame member and remove the two dome nuts and larger locknuts situated below them (if crash bars have already been removed from the machine, only the upper dome nut will need to be removed). Firmly grip both ends of the removable member and pull it clear of the machine.

19 Move to the right-hand side of the machine and remove the

lower mid-point mounting bolt and nut, followed by the lower forward locknut and washer (if not already removed with the crash bars). The lower forward mounting stud may now be driven out of its location. Note that it may be necessary to adjust the pressure applied by the jack to the engine sump to facilitate removal of the bolt and stud.

20 The engine will now be suspended in the frame by the fan shroud mounting bolts and the two rear upper mounting plates. Before disturbing any of these attachment points, unclip all the electrical leads from the fan shroud and check around the engine/gearbox unit to ensure all electrical leads, control cables, breather pipes, etc, are clear of the unit. Remove the bolt passing through the two rear upper mounting plates and detach the plates from the frame by removing the remaining flange bolts. At this stage it is necessary to have an assistant steadying the engine unit. Remove the four bolts which secure the fan shroud to the front upper engine mounting lugs and to the frame. With assistants supporting the engine unit, manoeuvre it sideways to the left slightly to allow the fan shroud to be pulled away from the machine. Check that the push-fit lead to the electric fan sensor switch is disconnected and proceed to carefully wheel the trolley jack and engine unit outwards to the left. If the throttle cables have not been disconnected previously, then this should be accomplished before the engine unit is moved further. Continue to move the engine outwards from the frame, continually checking to see if any part of the engine is fouling the frame or fuel tank. It may be found necessary to raise or lower the engine unit slightly or to pivot it around the jack head in order to make it clear the frame. Note that the vacuum diaphragm casing of the CDI pulser generator will catch on the left-hand rear downtube of the frame unless care is taken. Once the engine unit is completely clear of the frame it can be lifted clear of the jack and placed securely on the work surface. When storing the mounting assemblies, place the battery earth lead with the left-hand rear upper mounting plate so that its position is not forgotten when refitting.

6.3 Remove coolant and oil drain plugs

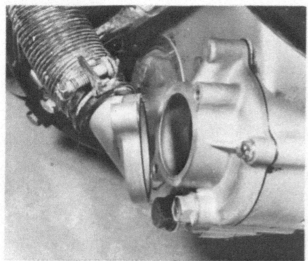

6.5 Remove the manifold from the water pump outer casing

6.10 Remove the tachometer drive cable retaining bolt from the fuel pump mounting casing

6.15a Remove the circlip from the engine shaft ...

6.15b ... to allow the universal joint to be detached

6.18a Remove the footrest retaining nut together with the frame member retaining nut

6.18b Locate and remove the engine casing to frame attachment bolt ...

6.18c ... together with the nuts from the frame member forward locating studs

6.20a Pull the fan shroud leftwards away from the machine

6.20b Move the engine unit out from the frame

7 Dismantling the engine/gearbox unit: preliminaries

1   Before any dismantling work is undertaken, the external surfaces of the unit should be thoroughly cleaned and degreased. This will prevent the contamination of the engine internals, and will also make working a lot easier and cleaner. A high flash point solvent, such as paraffin (kerosene) can be used, or better still, a proprietary engine degreaser such as Gunk. Use old paintbrushes and toothbrushes to work the solvent into the various recesses of the engine castings. Take care to exclude solvent or water from the electrical components and inlet and exhaust ports. The use of petrol (gasoline) as a cleaning medium should be avoided, because the vapour is potentially explosive and can be toxic if used in a confined space.

2   When clean and dry, arrange the unit on the workbench, leaving a suitable clear area for working. Gather a selection of small containers and plastic bags so that parts can be grouped together in an easily indefinable manner. Some paper and a pen should be on hand to permit notes to be made and labels attached where necessary. A supply of clean rag is also required.

3   An assortment of tools will be required, in addition to those supplied with the machine (see 'Working conditions and Tools' for details). Unlike owners of most Japanese machines, those working on the Honda GL 1100 do not suffer unduly from the normal soft-headed cross-point screws. In most areas, small hexagon-headed screws are employed, these being much more robust. In view of this, an appropriate 'nut driver' or box spanner will prove invaluable, in addition to the normal range of workshop tools.

4   Before commencing work, read through the appropriate section so that some idea of the necessary procedure can be gained. When removing the various engine components it should be noted that undue force is seldom required, unless specified. In many cases, a component's reluctance to be removed is indicative of an incorrect approach or removal method. If in any doubt, re-check with the text.

8 Dismantling the engine/gearbox unit : removing the carburettors

1   The carburettors and manifold assembly can be removed as a single unit, further dismantling only being necessary if the carburettors require attention (see Chapter 3).

2   Remove the fuel feed pipe to the rearmost right-hand carburettor at the fuel pump outlet stub and the feed pipe to the pulser generator at its diaphragm casing stub. The two domed bolts which retain each carburettor manifold to the cylinder head may now be removed and the complete assembly lifted upwards.

9 Dismantling the engine/gearbox unit : removing the thermostat housing

1   Remove the three screws which hold the thermostat housing to the crankcase, and the two screws which retain each transfer manifold. Lift the complete unit from position noting the hollow dowel and O-ring which locate the housing in position. It will take a fair degree of effort to separate the transfer manifolds from their cylinder head connections because the gaskets used are pre-treated with a sealing compound which acts as a very strong adhesive. Tapping the transfer manifolds with a soft-wood block will help to break the gasket seal. Needless to say, care should be exercised during this operation to avoid damage.

2   With the assembly placed on a clean work surface, pull the transfer pipes out of the housing and their manifolds. Note the four O-rings. If required, the hose manifold can be detached from the housing, after removing the two bolts, and the thermostat lifted from position.

8.2 The complete carburettor assembly may be lifted from the engine

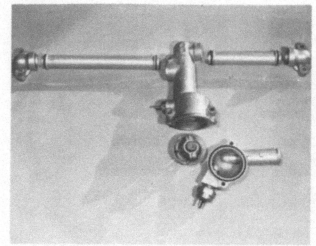

9.2 Dismantle the thermostat housing assembly on a clean work surface

10 Dismantling the engine/gearbix unit : removing the timing belts

1   Remove the left-hand timing belt cover followed by the right-hand cover. Each cover is held in position by two special bolts. Care must be taken to ensure that the cover seals part cleanly and do not split.

2   Remove the flywheel timing mark inspection plug, which is located on the upper rear portion of the crankcase, and remove the alternator cover cap to expose the centre bolt head. Rotate the crankshaft, using a spanner on the generator centre bolt, until the 'T' mark on the flywheel aligns with the marks cast either side of the inspection plug hole. To ensure the marks are accurately aligned, it may be easier to align the flywheel mark with a thin piece of stiff wire placed across the index marks on the orifice. Cut the wire to a suitable length so that it cannot be displaced easily and so fall into the crankcase.

3   With the engine set in this position check that the driven pulley timing marks are correctly positioned in alignment with the index marks on the inner case extensions. See Fig. 1.19. If the marks are not easy to see scribe one on each pulley to aid subsequent assembly. Similarly mark the pulleys left or right, so that on reassembly they may be fitted in their original positions.

4   Place a spanner on the drive pulley centre bolt and sharply tap the end of the spanner so that the bolt is jarred free. The weight of the reciprocating parts will prevent the engine from turning. Slacken off the jockey pulley bolts.

5   Before removing the timing belts, mark each with a piece of tape to indicate the original direction of travel. If a belt is subsequently fitted so that the direction of travel is reversed, belt wear will be accelerated.

6   Ease the outermost (right-hand) timing belt from position on the pulleys, lifting the jockey pulley against its spring to provide as much slack in the belt as possible. Remove the innermost belt using the same method. Considerable care should be exercised when removing the belts as they can easily become damaged. The belts are manufactured from synthetic rubber strengthened by glass fibre strands, and bending to a radius of less than 25.0 mm (1.0 in) or scoring with a screwdriver during removal, will considerably shorten their life. Avoid contact between the belts and engine oil or grease as the lubricant may swell the rubber, causing deterioration and maladjustment of valve timing.

7   Remove the centre bolts from the two driven pulleys and draw the pulleys from position. Note the Woodruff key in each camshaft, which should be removed using a suitable screw-driver. The pulleys can be prevented from rotating while the centre bolt is removed, by placing a tyre lever through the spokes, abutting against one of the bolt heads which lie behind the pulley. **Warning:** Do not allow the camshafts or the crank-shaft to rotate independently of each other as there is a danger that a piston may make contact with an open valve.

8   Unscrew the already loosened drive pulley centre bolt and washer and pull the guide plates and pulleys from position noting the position of the Woodruff keys. Reassemble the plates and pulleys in their correct order as they are removed and place the complete assembly to one side.

8   Remove the inner case extension from each cylinder head. Each is retained by two bolts.

11 Dismantling the engine/gearbox unit : removing the cylinder heads

1   Loosen and remove the two flange bolts that secure the fuel pump/tachometer drive assembly to the rear of the right-hand cylinder head and remove the assembly as a complete unit. Move to the rear of the left-hand cylinder head and remove the camshaft and sealing cap retaining cover and gasket by unscrewing the two flange bolts.

2   The cylinder heads should now be removed individually using an identical procedure, as follows. Unscrew the cylinder head/cam cover retaining bolts and lift off the cover. Place a small container or old rag below the cylinder head to catch the small amount of oil trapped in the cover. When removing the cover, take care not to damage the rubber sealing ring.

3   Loosen the seven cylinder head retaining bolts evenly and in a diagonal sequence. This will help prevent distortion. Note that in addition to the six 10 mm bolts, an extra 6 mm bolt is fitted on the lower flange of the cylinder head. The cylinder head gasket is of the impregnated type and consequently after a certain amount of service the head will become bonded to the cylinder barrel. A rawhide mallet may be used to free it, or in extreme cases, a block of soft wood placed on suitable portions of the head and tapped with a hammer may be utilised. Have an assistant support the cylinder head so that it does not fall free and become damaged. After removal of the head, pull the two hollow dowels from position together with the oil feed control connector. Note the two O-rings on the connector.

4   It should be noted that two of the 10 mm bolts fitted through each cylinder head are longer than the other four. Note their positions for reference when refitting.

10.1 Each timing belt cover is retained by two special bolts

10.2 Align the T mark on the flywheel with the marks either side of the inspection hole

10.3 Slacken the jockey pulley bolts

10.6 Mark the position of the driven pulleys in relation to the mark on the inner case extension

10.7 Note the order in which the drive pulleys and guide plates are removed

11.1 Remove the camshaft end sealing cap retaining cover

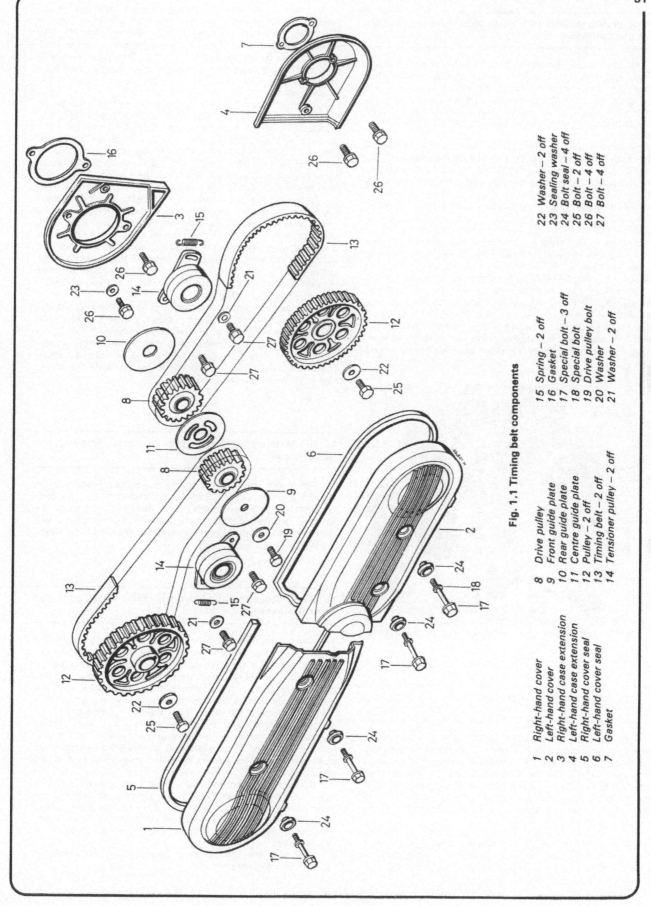

Fig. 1.1 Timing belt components

1 Right-hand cover
2 Left-hand cover
3 Right-hand case extension
4 Left-hand case extension
5 Right-hand cover seal
6 Left-hand cover seal
7 Gasket

8 Drive pulley
9 Front guide plate
10 Rear guide plate
11 Centre guide plate
12 Pulley – 2 off
13 Timing belt – 2 off
14 Tensioner pulley – 2 off

15 Spring – 2 off
16 Gasket
17 Special bolt – 3 off
18 Special bolt
19 Drive pulley bolt
20 Washer
21 Washer – 2 off

22 Washer – 2 off
23 Sealing washer
24 Bolt seal – 4 off
25 Bolt – 2 off
26 Bolt – 4 off
27 Bolt – 4 off

12 Dismantling the engine/gearbox unit : removing the CDI pulser generator assembly

1   The CDI pulser generator assembly is located on the rearmost end of the crankshaft. To gain access to the assembly remove the metal housing cover by unscrewing its three retaining flange bolts. Check that the cover gasket is in good condition and renew it if thought necessary. It is imperative that no contamination should be allowed to enter the generator housing.

2   The pulser generator assembly and its housing may now be removed by unscrewing the three hexagon-headed retaining bolts and pulling the housing off its locating pin on the engine casing. Unscrew and remove the automatic timing1 unit (ATU) retaining bolt and plain washer and pull the assembly off the crankshaft extension end. Lift the thick insulator gasket from the engine casing, taking care to guide it over the locating pin. Finally, remove the ATU locating pin from its recess in the crankshaft extension using a pair of long-nose pliers. Take care whilst doing this to avoid damaging the engine casing mating surfaces and seal.

3   Check the complete CDI pulser generator assembly for any signs of contamination and clean the components if necessary. If the assembly if fully serviceable, it should be wrapped in a clean piece of material and stored in a clean, dry space until required for reassembly. If suspected of being unserviceable, inspect the assembly as detailed in the relevant Sections of Chapter 4 and take the necessary steps to obtain any new parts required.

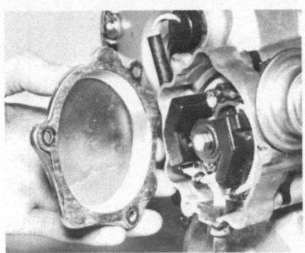

12.1 The CDI pulser generator assembly is located beneath the housing cover

13 Dismantling the engine/gearbox unit : removing the clutch

1   If required, the clutch cover, plates and springs can be removed with the engine in the frame. Commence by removing the four long and two short 6 mm cover securing bolts, noting their respective positions for reference when refitting, followed by the two 6 mm flange nuts. Place a small container or piece of rag beneath the cover and pull the cover away from the engine casing, allowing any oil contained within the clutch assembly to drain out.

2   Remove the short operating rod from within the bearing centre and place it in the clutch cover to avoid loss. Unscrew the six flange bolts which retain the clutch lifter plate in position, unscrewing them only in small increments and in a criss-cross pattern to avoid breaking the plate. Note that two of these bolts are longer than the other four and their position should be noted for reference when refitting; they may be readily identified when fitted by the different colour of their heads compared with the other bolts. The lifter plate, complete with bearing and operating rod guide, may now be lifted from position. Remove the clutch springs from their locations on the pressure plate.

3   The clutch centre is secured by a slotted nut, for which special Honda tools Nos. 07923 – 3710000 and 07716 – 0020202 are available. A suitable tool can, however, be made up in the workshop using a short length of thick-walled tubing. Clamp the tube in a vice amd use a hacksaw to cut slots as shown in Fig. 1.3. The shaded area can then be filed away to leave four projecting tangs. These can be used to engage the slots in the nut to facilitate removal.

4   To prevent the clutch centre from rotating whilst the slotted nut is being loosened, refit two or more clutch springs and bolts using suitable washers to take the place of the lifter plate. Sufficient pressure will be given to lock the two parts of the clutch together. Remove the inspection cap from the centre of the alternator cover and apply a spanner to the generator rotor centre bolt. The clutch centre nut can now be removed together with the special washer. Note this washer is marked OUTSIDE for reference when refitting.

5   Remove the refitted clutch springs and bolts and pull out the clutch centre followed by the clutch plates. Carefully note the sequence of plates to aid refitting, noting particularly the position of the central damper plate and the outermost smaller width friction plate. Finally, remove the internal splined washer. The clutch outer drum must remain in position until the engine rear cover has been removed.

14 Dismantling the engine/gearbox unit: removing the starter motor

1   The starter motor is retained by two bolts passing through lugs on the motor body. Remove the bolts and using a soft-faced mallet, gently tap the starter motor from the casing. The sprocket, with which the starter splined shaft engages, will remain in situ within the rear cover, still meshed with the chain.

15 Dismantling the engine/gearbox unit : removing the rear cover and clutch outer drum

1   Loosen and remove the twelve rear cover flange bolts. The bolts should be slackened evenly and in a diagonal sequence to avoid distortion of the cover casting. Remove the bolts, noting the position of the electrical lead clamp and the clutch cable guide for reference when refitting. Note also that one of the twelve flange bolts is of a shorter length than the others, this being the lower of the clutch cable guide retaining bolts.

2   Remove the cover by pulling it from the two locating dowels. It may be found necessary to tap the cover with a soft-faced mallet in order to help free it. When removing the case, push the final drive splined shaft inwards, so that it is not disturbed.

3   Remove the Allen bolt and washer from the end of the oil pump drive shaft to free the oil pump sprocket, and remove the clutch outer drum retaining circlip. Ease the oil pump driven sprocket and the clutch outer drum off their respective shafts simultaneously. The sprockets can then be detached from the duplex chain.

4   Withdraw the CDI pulser generator drive shaft from its location in the crankshaft end.

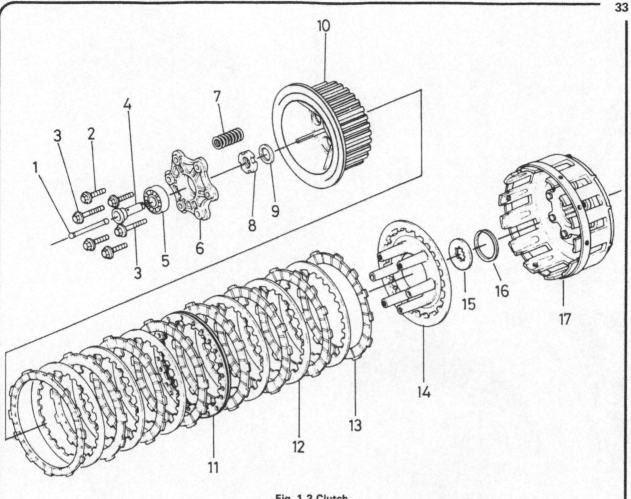

Fig. 1.2 Clutch

1 Operating rod	7 Spring – 6 off	13 Friction plate – 8 off
2 Bolt – 4 off	8 Slotted nut	14 Pressure plate – 6 off
3 Bolt – 2 off	9 Special washer	15 Splined washer
4 Operating rod guide	10 Clutch centre	16 Washer
5 Bearing	11 Damper plate	17 Outer drum
6 Lifter plate	12 Plain plate	

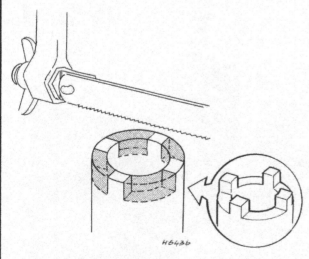

H6436

Fig. 1.3 Home-made clutch nut peg spanner

13.1 Remove the clutch housing cover together with operating mechanism

13.2 Note the lengths and fitted positions of the clutch lifter plate retaining bolts

13.4a Prevent the clutch centre from rotating ...

13.4b ... before loosening and removing the slotted nut

13.4c The special washer should be refitted with the marked face facing the nut

14.1 Disengage the starter motor from the starter drive sprocket

15.3 Remove the oil pump sprocket retaining bolt and washer

16 Dismantling the engine/gearbox unit : removing the alternator rotor and the starter clutch assembly

1   Loosen and remove the alternator rotor retaining bolt and washer. The rotor may be prevented from turning by inserting a tommy bar of the correct diameter into one of the balance holes drilled in the starter clutch housing. The rotor and starter clutch assembly, to which it is attached, can now be drawn off the splined shaft. It is likely that the assembly will be tight on the splines, in which case an extractor must be used for removal. A standard two or three legged sprocket puller may be employed. Do not try prising the assembly from position using screwdrivers or other tools as levers. The mating sufaces of the casing and the rear face of the starter clutch will probably suffer damage. Be prepared to take the weight of the alternator rotor as it leaves the shaft. It is a heavy component weighing 9 lbs.
2   Remove the splined washer from the alternator shaft. Loosen the chain guide retainer bolt and rotate the guide so that it clears the chain. Pull the sprocket off the shaft together with the chain and starter drive sprocket.

17 Dismantling the engine/gearbox unit : removing the final drive shaft

1   Remove the four bolts which secure the final gear cover to the engine casing. This cover is located just to the rear of the right-hand cylinder head. The gasket located beneath the cover will now have to be peeled off the engine casing mating face so that the final drive double gear is exposed.
2   Pull the drive shaft rearwards out of the engine casing. This will allow the final drive double gear to be rolled out of the opening to the rear of the cylinder head.

18 Dismantling the engine/gearbox unit : removing the water pump and front engine cover

1   Remove the four screws which retain the water pump outer casing. The pump casing is located on two dowels and fitted with a self-sealing gasket; it may be tight and require easing with a soft-faced mallet. Remove the nine screws which retain the front cover and lift the cover from position, noting the position of the electrical wire retaining clip located beneath one of the screw heads. Note also that the screws are of varying lengths; their positions should be noted for reference when

refitting by inserting them in a cardboard template of the cover. Remove the cover gasket and note the position of the O-rings (including those around the two collars and locating dowel).
2   The water pump is retained in the front cover, from the inside by three bolts. After removal of the bolts, tap the pump body gently until it leaves the casing. Do not tap the pump shaft as damage may result. Note the two O-rings which fit in locating grooves around the double diameter periphery of the pump body.

19 Dismantling the engine/gearbox unit : separating the crankcase halves

1   Remove the three crankcase half holding bolts from the right-hand side of the engine. One bolt is positioned at the upper rear corner and two at the lower rear corner. Place the engine so that the right-hand cylinder block mating surface rests securely on the workbench. Disengage the spring loaded change pawl, which pivots on the gearchange arm, from the change drum pins. Tie the pawl back to the change arm so that it will not snag any part of the gear change mechanism in the right-hand casing.
2   Loosen the nineteen left-hand crankcase bolts evenly and in a diagonal sequence. Remove all the bolts. The bolts are of different lengths and should be inserted into a cardboard template so that their respective positions are noted for reference when refitting. Note also that the long bolt located to the rear of the left-hand cylinder barrels has a special copper sealing washer fitted beneath its head.
3   Separation of the crankcase halves requires two people. One person to lift the left-hand casing upwards and the second person to support the pistons, preventing them from falling free and becoming damaged as they leave the cylinder bores. Use a soft-faced mallet to loosen the cases. Some difficulty will be experienced in freeing the cases from their locating dowels. Resist the temptation to insert a lever between the casing mating surfaces, the most efficient way found to free the cases was by placing a wood block on a projection on the upper casing and tapping it sharply upwards with a heavy hammer. This process was repeated on various projections around the casing and eventually the casing was freed.
4   Once clear of the dowels, lift the left-hand case upwards so that it remains square in all planes and the pistons do not tie in the cylinder bores. Broken piston rings will fall into the right-hand crankcase from where they should be retrieved immediately. Place a clean rag over the upper edge of the right-hand casing and rest the two pistons upon it.

16.1 Loosen and remove the alternator rotor retaining bolt and washer

16.2a Loosen the chain guide retainer bolt ...

16.2b ... and remove the sprockets and chain

17.1 Remove the final gear cover ...

17.2 ... to allow the final drive double gear to be rolled out of position

19.1 Tie the spring loaded change pawl to the change arm

19.2 Note the partially hidden crankcase bolt

19.4 Keep the left-hand case square in all planes whilst lifting

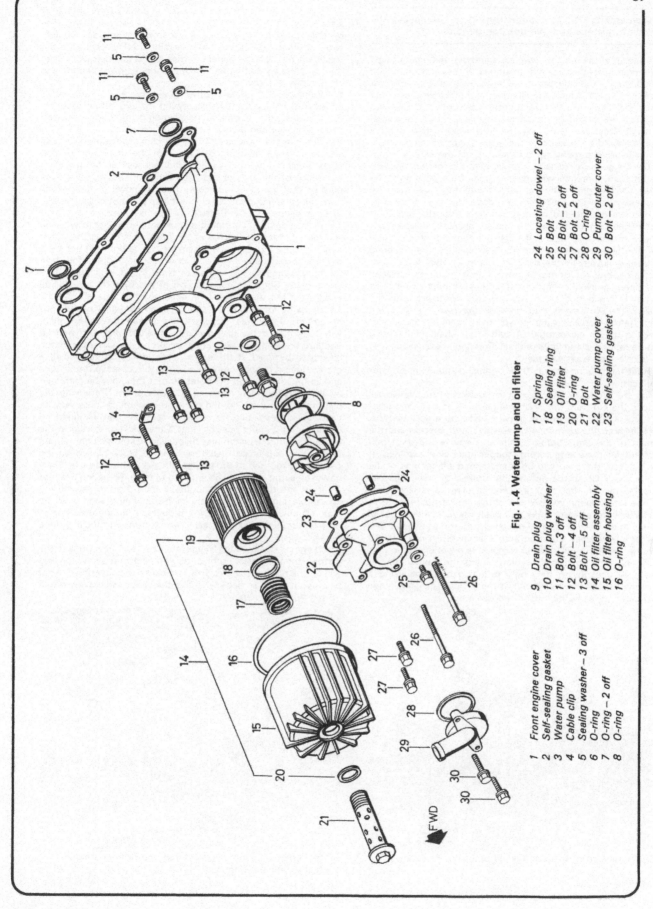

Fig. 1.4 Water pump and oil filter

1 Front engine cover
2 Self-sealing gasket
3 Water pump
4 Cable clip
5 Sealing washer – 3 off
6 O-ring
7 O-ring – 2 off
8 O-ring

9 Drain plug
10 Drain plug washer
11 Bolt – 3 off
12 Bolt – 4 off
13 Bolt – 5 off
14 Oil filter assembly
15 Oil filter housing
16 O-ring

17 Spring
18 Sealing ring
19 Oil filter
20 O-ring
21 Bolt
22 Water pump cover
23 Self-sealing gasket

24 Locating dowel – 2 off
25 Bolt
26 Bolt – 2 off
27 Bolt – 2 off
28 O-ring
29 Pump outer cover
30 Bolt – 2 off

FWD

20 Dismantling the engine/gearbox unit: removing the crankshaft, pistons and gearbox components

1   Check that the main bearing caps and big-end bearing caps are marked in relation to their positions. If they have not been marked, this should be accomplished now, using a punch or scribe, so that no confusion will arise on replacement as to the correct positions. Do not obliterate the scribed letters and numbers already marked, as these are code numbers for journal and shell sizes, which will be required when ordering spare parts. It should be noted that the arrows marked on the main bearing caps must point towards the top of the engine.
2   Tilt the gearbox mainshaft upwards at the forward end and pull the shaft complete with gear pinions from position in the primary driven gear. The primary driven gear can now be disengaged from the Hy-vo primary drive chain.
3   Remove the big-end bearing cap nuts from the two right-hand connecting rods, rotating the crankshaft as necessary. Remove the bearing caps and push the connecting rods away from the big-end bearing journals. Remove the main bearing cap bolts followed by the caps. If a bearing shell falls out of either a big-end cap or main bearing cap it should be replaced immediately to avoid confusion. Grasp the left-hand pistons and the Hy-vo chain and lift them upwards complete with the crankshaft. Take care not to damage the piston rings when lifting. Before displacing the Hy-vo chain, mark its position for reference during reassembly. A part-worn chain of this type must be refitted to run in its original direction or excessive noise and accelerated wear will result.
4   Refit the bearing caps in their original positions to aid later identification.
5   Lift the oil gauze filter box from position in the crankcase. Remove the nut from the gear change drum stopper arm pivot shaft and remove the two stopper arms and the spacer washers. Carefully note the sequence of washers and spacers and the position of the arms before removal. Remove the pivot bolt which retains the change drum stopper inner claw assembly. If care is taken, the claw can be manoeuvred off the end of the change drum. Grasp the end of the retainer pin which passes vertically through the crankcase and locates the selector fork rod. The retaining pin will pull out of the case. Note the relative positions of the selector forks and push the selector rod out of position in the case so that all three selector forks are freed. The selector fork rod has a slotted end. A screwdriver can be used to rotate the rod if difficulty in removal is experienced.
6   Remove the small plate whicn retains the neutral stopper switch, by unscrewing its single retaining bolt, and using a screwdriver, push the switch out of the casing. The gearchange drum can now be pulled out of the casing.

7   Using an 'impact' screwdriver remove the two countersunk screws which retain the blind layshaft bearing housing. The bearing housing is a push-fit and can be removed, complete with the bearing. Pull the 5th gear pinion off the layshaft and remove it through the bearing orifice in the gearbox wall. Slide the complete layshaft assembly out of the rearmost bearing and lift it out of the casing.
8   Remove the primary drive chain guide, if renewal is considered necessary, by unscrewing its two retaining bolts. Pull the oil supply nozzle from its location in the rearmost casing mating surface and store it in a clean, dry place.
9   Remove the various bearing half-clips from their locations and remove the locating dowels from the crankcase mating faces. Position the crankcase so that access can be gained to both ends of the pistons and push each connecting rod up the bore so that the pistons become free. Removal of the pistons from the connecting rods should not take place at this stage due to the reasons given in Section 27 of this Chapter.
10   If the engine has covered a substantial mileage, a wear lip will have formed at the top of the bore. This marks the upper limit of piston top ring travel. Remove the carbon that will have built up above the wear lip and ascertain the degree of wear in the bore. Compare the degree of wear found with that figure given in the Specifications Section of this Chapter and decide whether or not the engine is in need of a rebore.
11   It will be found that the piston cannot be easily pushed out of the bore if any wear lip exists. The degree of wear however, will determine if the piston can be pushed past the wear lip without breakage of the rings occurring or excessive force having to be used. If the engine is found to be in need of a rebore, then it may be considered advisable to leave the pistons and connecting rods in the bores so that they may be extracted by the Service Agent carrying out the rebore. Removing the piston past the wear lip and deciding how much force may be used to achieve this, is really a task for the experienced mechanic. It is quite possible that if the degree of wear in the bores is slight, the piston may be pushed past it with only a light tap from a soft-faced mallet on a wood block placed on the connecting rod end. If the degree of wear in the bores is heavy however, some force will be required to push the piston past the wear lip and this will certainly break the piston rings with the accompanying risk of the bore or piston being scored or the piston breaking around the ring grooves. If in any doubt as to how to complete this task, seek advice from an official Honda Service Agent.
12   It should be noted that if it is intended to fit new rings to the pistons, then the wear lip must be removed as the new top ring will hit the ridge and break. This is because the new ring will not have worn in the same way as the old, which will have worn in unison with the ridge.

20.3 Lift the crankcase assembly out of the engine casing

20.5a Remove the nut (arrowed) from the stopper arm pivot shaft and detach the arms ...

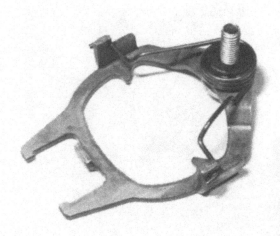

20.5b ... followed by the inner claw assembly

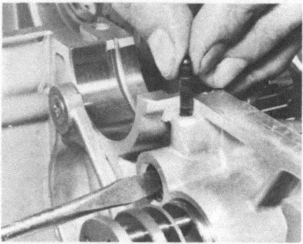

20.5c Withdraw the selector rod retaining pin

20.6 The neutral stopper switch is retained in position by a plate and bolt

20.8 Inspect the primary drive chain guide

21 Dismantling the engine/gearbox unit : removing the oil pumps, alternator shaft and gearchange shaft

1   All the components listed in the Section heading are located in the left-hand crankcase half.

2   Remove the Hy-vo chain tensioner, which is retained by two bolts, by first lifting the tensioner plate so that a socket or box spanner may be fitted over the bolt head nearest the spring. Remove this bolt and lower the tensioner plate so that the same socket may be passed through the access hole to the remaining bolt head. With both bolts removed, the tensioner assembly may be lifted out of the casing.

3   Detach the small oil guide plate from the crankcase wall by unscrewing its two retaining bolts. The alternator shaft, complete with pinion and shock absorber, is retained by a detachable bearing housing. Remove the five retaining bolts and pull the housing, complete with shaft, from position. Note that one of the five bolts is longer than the others and that the chain guide plate is located under the head of the lower bolt.

4   The oil pumps are fitted at either end of the crankcase, being driven by a shaft which passes through and between the two. Detach the clutch scavenge pump, which is the smallest of the two and retained by three bolts. Pull the pump off the shaft. Some difficulty may be experienced in removing the pump from the shaft; it was found to be a very tight fit. The main pump can

now be removed in a similar manner, together with the drive shaft, which will remain connected. Note the two hollow dowels on which the pump locates.

5   Remove the gearchange shaft ball arm retaining bolt, after knocking down the ears of the tabwasher, and pull the shaft out of the casing. Note the shaft collar and the positioning of the return spring. The ball arm may now be detached from the gearchange pedal shaft and the shaft withdrawn from the casing. The return spring locating pin may be left in position.

6   Remove the one remaining oil catch plate retaining bolt and lift the plate from position, gearbox end first.

22 Examination and renovation : general

1   Before examining the parts of the dismantled engine unit for wear it is essential that they should be cleaned thoroughly. Use a petrol/paraffin mix or a high flash-point solvent to remove all traces of old oil and sludge which may have accumulated within the engine.

2   Examine the crankshaft castings for cracks or other signs of damage. If a crack is discovered it will require a specialist repair.

3   Examine carefully each part to determine the extent of wear, checking with the tolerance figures listed in the Specifications section of this Chapter. If there is any question of doubt play safe and renew.

4   Use a clean lint free rag for cleaning and drying the various components. This will obviate the risk of small particles obstructing the internal oilways, and causing the lubrication system to fail.

5   Various instruments for measuring wear are required, including a vernier gauge or external micrometer and a set of standard feeler gauges. Honda recommend the use of Plastigage for measureing radial clearances between working surfaces such as shell bearings and their journals. Plastigage consists of a fine strand of plastic material manufactured to an accurate diameter. A short length of Plastigage is placed between the two surfaces, the clearance of which is to be measured. The surfaces are assembled in their normal working positions and the securing nuts or bolts fastened to the correct torque loading; the surfaces are then separated. The amount of compression to which the gauge material is subjected and the resultant spreading indicates the clearance. This is measured directly, across the width of the Plastigage, using a pre-marked indicator supplied with the Plastigage kit. If Plastigage is not available both an internal and external micrometer will be required to check wear limits. Additionally, although not absolutely necessary, a dial gauge and mounting bracket is invaluable for accurate measurement of end float, and play between components of very low diameter bores – where a micrometer cannot reach.

21.2 The Hy-vo chain tensioner is retained in position by two bolts

21.3a Detach the oil guide plate from the crankcase wall

21.3b Remove the alternator shaft, complete with pinions and shock absorber

21.5a Remove the gearchange shaft ball arm retaining bolt ...

21.5b ... and pull the change arm shaft out of the casing ...

21.5c ... followed by the pedal shaft

21.6 Lift the oil catch plate from position

23 Big-end bearings : examination and renovation

1 Big-end failure is invariably indicated by a pronounced knock from the crankcase. The knock will become progressively worse, accompanied by vibration. It is essential that the bearings are renewed as soon as possible since the oil pressure will be reduced and damage caused to other parts of the engine.
2 Bearing wear at the big-ends can only be really accurately assessed after separating the crankcase halves. A rough indication can, however, be obtained after removal of the cylinder head in the following manner. Starting on No one (1) cylinder, rotate the engine so that the piston reaches TDC and then starts to move down the bore. Stop the crankshaft. In this position the big-end journal is pulling away from the connecting rod. Using two thumbs press down sharply on the piston crown. Any movement of the piston which is not accompanied by movement of the crankshaft indicates that wear has developed either at the big-end or between the working surfaces of the gudgeon pin and piston bosses. In either case the crankcases will have to be separated for further examination and work.
3 The big-ends have shell type bearings. Examine the bearing surfaces after removing the bearing caps; it is not necessary to remove the shell. If the bearings are badly scuffed or scored the shells will have to be renewed. Always renew bearings as a complete set. If the shell surfaces are excessively scuffed or have 'picked-up' and the journal on that bearing is 'blued' or discoloured, it may be an indication of lubrication failure. The functioning of the lubrication system MUST be checked before engine reassembly in such a case.
4 If the condition of the bearings appears to be satisfactory, check that the clearance between each bearing and the journal is within the recommended limit as follows:

Standard clearance   0.020 – 0.044 mm
                               (0.0008 – 0.0017 in)
Service limit   0.080 mm (0.0031 in)

The clearance can be assessed by measuring the internal diameter of the shell bearing and the outside diameter of the big-end journal and the outside diameter of the big-end journal and subtracting the second figure obtained from the first. 'Plastigage' can also be used in the following manner. Cut a short length of 'Plastigage' and place it on the journal so it lies along the line of the bearing axis. Bolt up the connecting rod and bearing cap to the recommended torque of 3.0 – 3.4 kgf m (22 – 25 lbf ft). Do not rotate the bearing. Separate the two bearing halves and measure the amount of spread of the 'Plastigage' using the gauge provided. This will indicate the amount of clearance.

5 When renewing the bearings, bearing selection should be made by referring to the connecting rod code (a number scribed on the machined side of the connecting rod) and to the crankpin code (a letter scribed on the adjacent crankshaft web). Cross-refer both letter and number to the accompanying selection chart.

Crank-pin code	Connecting rod code No. / Connecting rod I. D. / Crankpin O. D.	1 46.000–46.008	2 46.008–46.016	3 46.016–46.024
A	42.992–43.000	Yellow	Green	Brown
B	42.984–42.992	Green	Brown	Black
C	42.976–42.984	Brown	Black	Blue

6 When fitting new bearings, ensure that they are positioned correctly and that the tongues on the end of each shell locate with the recesses in the connecting rod or bearing cap. Also check the clearance on each bearing to ensure that selection is accurate. It is considered good practice to renew all bearing shells when an engine is stripped down, irrespective of their condition. Shell bearings are relatively inexpensive compared with the cost of subsequent dismantling to replace bearings that have failed prematurely.

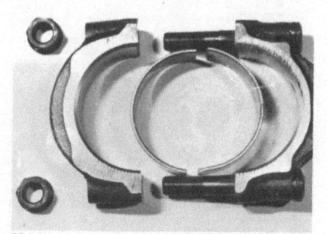

23.3 Remove the big-end bearing cap and examine the bearing shells

23.4 Using the scale supplied to measure the spread of Plastigage

23.5 Bearing selection should be made by referring to the code number (letter indicates weight code)

23.6 Locate the tongue on each bearing shells with the cap recess (arrowed)

24 Crankshaft and main bearings : examination and renovations

1   Examine the main bearing shells and check the clearances, using the procedure as described for big-end bearings. The correct clearances are as follows:

Standard clearance	*0.020 – 0.044 mm (0.0008 – 0.0017 in)*
Service limit	*0.080 mm (0.0031 in)*

2   Measure the main bearing and big-end bearing journals for ovality by measuring each journal in several radial positions, with a micrometer. If any journal is worn beyond the service limit, the crankshaft must be renewed.

Standard ovality	*0.005 mm (0.0002 in)*
Service limit	*0.008 mm (0.0003 in)*

Check each journal for taper over the full width. If the taper exceeds the service limit the crankshaft must be renewed.

Standard taper	*0.002 mm (0.00008 in)*
Service limit	*0.004 mm (0.0002 in)*

3   Support the crankshaft at each end on vee-blocks or between centres. Rotate the crankshaft and measure the runout at the centre main bearing journal, using a dial gauge. If the runout exceeds the service limit, the crankshaft must be renewed, or placed in the hands of a specialist for straightening.

Maximum runout	*0.05 mm (0.0019 in)*

Bear in mind that the runout is half the actual reading taken.
4   Main bearing shells must be selected by referring to the code numbers scribed on the adjacent crankshaft webs and by the main bearing cap code numbers or letters which are marked on the crankcase. The coding on the crankcase can be found behind the right-hand cam belt jockey pulley. The Arabic numerals refer to the respective main bearings. The Roman numerals or capital letters indicate the code number. Cross refer the main bearing journal numbers and the crankcase code numbers with the accompanying chart for correct bearing selection.

Crankshaft journal code	Crankcase code No. / Main bearing I. D. / Crankshaft journal O. D.	I 52.000–52.008	II 52.008–52.016	III 52.016–52.024
1	47.992–48.000	Yellow	Green	Brown
2	47.984–47.992	Green	Brown	Black
3	47.976–47.984	Brown	Black	Blue

5   Check the security of the ball bearings which are used to plug the oilways in the crankshaft. They occasionally work loose, causing lubrication problems, and subsequent bearing failure. A loose ball can be carefully caulked back into place.

25 Connecting rods : examination and renovation

1   It is unlikely that a connecting rod will bend during normal usage, unless an unusual occurrence such as a dropped valve has caused the engine to lock. It is not advisable to straighten a connecting rod; renewal is the only satisfactory solution.
2   The connecting rods do not have small-end bearings in the accepted sense. Each gudgeon pin is a high interference fit in

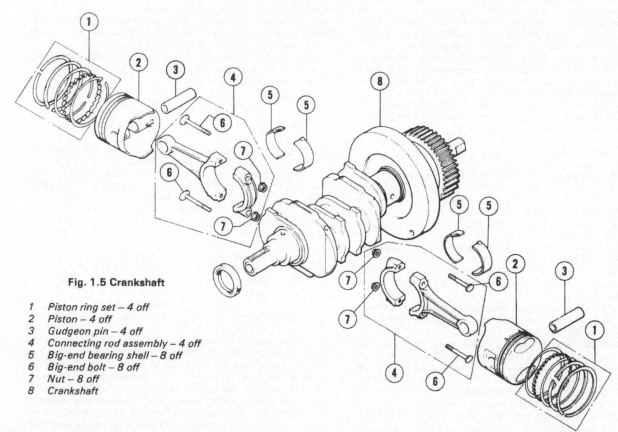

Fig. 1.5 Crankshaft

1   *Piston ring set – 4 off*
2   *Piston – 4 off*
3   *Gudgeon pin – 4 off*
4   *Connecting rod assembly – 4 off*
5   *Big-end bearing shell – 8 off*
6   *Big-end bolt – 8 off*
7   *Nut – 8 off*
8   *Crankshaft*

the small-end eye, the small-end bearing surfaces being in the piston bosses.

3   When checking big-end bearing clearances also check the side play of each connecting rod; using a feeler gauge. The correct clearances are as follows:

Standard axial clearance	*0.15 – 0.30 mm*
	(0.0059 – 0.0118 in)
Service limit	*0.40 mm (0.0158 in)*

If the clearance on any connecting rod exceeds the wear limit that rod must be renewed.

4   If, for any reason, it is found necessary to renew a connecting rod, the weight code of the replacement rod must be checked with that of the unserviceable item. The weight code is the letter scribed on the machined side of the connecting rod, just below the connecting rod code number. A new connecting rod should be selected with a weight code the same as that of the old rod.

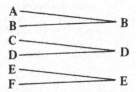

Fig. 1.6 Connecting rod weight code table

26 Cylinder bores : examination and renovation

1   The usual indication of badly worn cylinder bores and pistons is excessive smoking from the exhausts and piston

'slap', a metallic rattle that occurs when there is little or no load on the engine. If the top of each cylinder bore is examined carefully, it will be found that there is a ridge on the thrust side, the depth of which will vary according to the wear that has taken place. This marks the limit of travel of the top piston ring.

2   As described in Section 20 this ridge should be removed from the right-hand cylinder bores before piston removal, if the extent of wear is sufficient to cause damage to the piston rings.

3   Using an internal micrometer, measure each bore for wear. Take measurements at a point just below the upper ridge, at the centre of the bore and about 1 inch from the lower edge of the bore. Take two measurements at each point, one at 90° to the other. If the diameter at any point exceeds the service limit, the cylinders must be rebored and a set of oversized pistons fitted.

Cylinder bore ID	*75.0 – 75.015 mm*
	(2.9528 – 2.9533 in)
Service limit	*75.10 mm (2.9567 in)*

Again measure each bore and check for taper over the maximum piston travel. Rebore if the taper on any cylinder exceeds the service limit.

Cylinder bore	*0.007 – 0.012 mm*
taper	*(0.0003 – 0.0004 in)*
Service limit	*0.05 mm (0.002 in)*

Check for ovailty by measuring in the manner described for checking bore size. If ovality exceeds the wear limit, a rebore is necessary.

Maximum bore ovality	*0.15 mm (0.0059 in)*

4   Honda supply pistons in four oversizes : 0.25 mm (0.010 in), 0.50 mm (0.020 in), 0.75 mm (0.030 in) and 1.0 mm (0.040 in).

5   Assuming that the checks detailed above have been carried

out and satisfactory results obtained, check each bore visually. Inspect for score marks or other damage that may have resulted from an earlier engine seizure or displaced gudgeon pins. A rebore will be necessary to remove any deep scores.

6   Check the cylinder block upper mating surface for warpage. This can be done using a ground surface plate or bar if one is available, or by using a sheet of plate glass. Do not use ordinary window glass as this has an unground surface. Plate glass is of the type used in old mirrors and as a surface for some table tops. Use a feeler gauge between the cylinder block and surface plate to measure the amount of warpage.

   Maximum warpage   *0.1 mm (0.0039 in)*

If the warpage exceeds that stated above, the cylinder block surface will require surface grinding. This is a specialised operation. The crankcases should be returned to a Honda service agent for renovation.

7   If the cylinder bores do not require attention but new piston rings are to be fitted (as is usual after engine dismantling), it is advised that each cylinder is 'glazebusted' before reassembly. This will increase the rate at which the new rings bed in, increasing compression, and will also improve cooling. 'Glazebusting' is a specialised operation and should be entrusted to an expert.

7   Note that the wear lip must be removed from the top of each bore, as the new top ring is unstepped and will therefore hit the wear lip ridge and break.

8   Inspect the coolant cavities in the water jacket of each cylinder bank. 'Furring up' around the transfer orifices in the mating surfaces can be removed by the careful use of a scraper. Heavy deposits within the jacket can most easily be removed after the engine has been reassembled and refitted into the frame, using a special flushing fluid or compound added to the coolant. Ensure that the type used is suitable for aluminium castings.

27 Pistons and piston rings : examination and renovation

1   If a rebore is necessary, ignore this Section where reference is made to piston and ring examination since new components will be fitted.

2   If a rebore is not considered necessary, examine each piston closely. Reject pistons that are scored or badly discoloured as the result of exhaust gases by passing the rings. Remove each ring either by carefully opening each ring using the thumbs, or by placing three thin strips of tin between the ring being removed and the piston (see illustration). The oil scraper ring comprises a special crimped ring 'sandwiched' between two thin plain rings. Special care must be taken when removing this ring.

3   Remove all carbon from the piston crowns using a blunt instrument which will not damage the surface of the piston. A hardwood wedge or strip of aluminium alloy are suitable tools. Clean away all carbon deposits from the valve cutaways and finish off with metal polish to produce a smooth shiny surface. Carbon will not adhere so readily to a polished surface.

4   Measure each piston ring groove width, which should be as follows:

Top and second ring	*1.205 – 1.220 mm*
	(0.0475 – 0.0480 in)
Service limit	*1,30 mm (0.0512 in)*
Oil ring	*2.505 – 2.520 mm*
	(0.0986 – 0.0992 in)
Service limit	*2.60 mm (0.1024 in)*

5   Generally, when the engine is stripped down completely, the piston rings are renewed as a matter of course unless the rings have only been fitted for a short time. If ring life is such that renewal is not warranted automatically, the rings should be

examined as follows. Check that there is no build up of carbon on the inside surface of the rings or in the ring grooves of the piston. Any build up should be removed by careful scraping. An old broken ring, the end of which has been ground to a chisel profile, is useful for this. Refit each ring in its respective groove and measure the ring side play using a feeler gauge. The clearances should be as follows:

Top and second ring	*0.015 – 0.045 mm*
	(0.0006 – 0.0018 in)
Service limit	*0.120 mm (0.0047 in)*

There is no measureable side clearance on the oil ring as the crimped ring spring loads the plain rings.

6   Place each ring into the cylinder bore separately and measure the end gap. Push the ring down from the top of the bore, using the piston skirt, so that the ring remains square to the bore and is positioned about $1\frac{1}{2}$ inches from the top. If the end gap exceeds the wear limit on any ring, the rings should be renewed as a set.

Top and second ring	*0.1 – 0.3 mm*
	(0.004 – 0.012 in)
Service limit	*0.6 mm (0.02 in)*
Oil ring side rail	*0.2 – 0.9 mm*
	(0.008 – 0.035 in)
Service limit	*1.10 mm (0.04 in)*

7   Unlike most motorcycle pistons, those fitted to the GL 1100 have the bearing surface of each gudgeon pin in the two piston bosses and not as is usual in the small-end eye or bearing. The gudgeon pin is a high interference fit in the small-end eye. Removal and replacement requires the use of a special press and guide mandrel. When piston renewal is necessary or if excessive play has developed between the gudgeon pin and piston bosses, the pistons complete with connecting rods should be returned by a Honda repair agent for inspection and replacement. The correct clearance between the gudgeon pin and piston bosses can only be checked accurately after pin removal.

8   Check the outside diameter of each piston at the skirt by taking a measurement at 90° to the gudgeon pin. If the piston is worn to below the service limit it must be renewed.

Piston skirt OD	*74.950 – 74.980 mm*
	(2.9508 – 2.9520 in)
Service limit	*74.850 mm (2.9468 in)*

Place each piston in its cylinder bore and using a feeler gauge, check that the clearance is within the specified limits.

Cylinder to piston clearance 0.020 – 0.065 mm	
	(0.0008 – 0.0026 in)
Service limit	*0.25 mm (0.0098 in)*

9   The piston crowns will show whether the engine has been rebored on some previous occasion. All oversize pistons have the rebore size stamped on the crown. This information is essential when ordering replacement piston rings or in reboring.

10 The letter stamped on each piston ring, indicates the top of the ring. It is essential that each ring be fitted so that it is the correct way up.

28 Cylinder head and valves : dismantling, examination, renovation and reassembly

1   Before examination of each cylinder head can be carried out, the camshaft must be removed. Do this by removing the six bolts which retain the rocker arm/spindle support and lift the

support from position. The bolts should be loosened a little at a time in a diagonal sequence to prevent the camshaft being strained by the unequally opened valves. The support is located on two hollow dowels which should be removed to avoid loss. Removal of the support will free the camshaft, which can be lifted out of the half bearings, together with the oil seals.

2   Remove any carbon from the combustion chambers before removing the valves. Use a blunt scraper, which will not damage the surface and finish with metal polish.

3   A valve spring compressor is required to remove the valves. Compress the springs and remove the valve collets. Relax the springs and remove the valve spring collar, valve springs, valve stem seal, valve spring seat and finally, the valve itself. Take care to keep these components together in a set because each valve assembly should be refitted in its original location. Any carbon in the inlet and exhaust ports and on the valve heads should now be removed by using the same method as described for the combustion chambers.

4   Before attending to the valve seats, check the valve guides and stems for wear. Valve clearance may be determined by subtracting the valve stem diameter from the internal diameter of the guide.

Inlet valve stem diameter	6.58 – 6.59 mm
	(0.2591 – 0.2595 in)
Exhaust valve stem diameter	6.55 – 6.56 mm
	(0.2579 – 0.2583 in)
Inlet and exhaust guide ID	6.60 – 6.62 mm
	(0.2598 – 0.2606 in)
Service limit	6.64 mm (0.2614 in)
Inlet valve to guide clearance	0.01 – 0.04 mm
	(0.0004 – 0.0016 in)
Service limit	0.08 mm (0.0032 in)
Exhaust valve to guide clearance	0.05 – 0.07 mm
	(0.0020 – 0.0028 in)
Service limit	0.10 mm (0.0039 in)

If a valve or guide exceeds the stated wear limit it must be renewed. If the valve/guide clearance is excessive but the valve stem is still within the limits and is in good condition, the guide alone may be renewed. The guides may be drifted from position using a double diameter drift of the correct diameter after the cylinder head has been heated to 80° – 100°C. The cylinder head must be heated evenly in an oven; heating using a blowtorch will cause local hotspots resulting in a warped casting. Prior to drifting out an old guide ensure that all carbon deposits on the port side of the guide have been scraped off. This will aid removal and prevent broaching of the cylinder head material. Fit the new guides using the same drift and with the cylinder heat at a similar temperature. A new guide will require reaming in order to bring the valve/guide clearance within the specified tolerances. It is probable also that the valve seat will require cutting or order that the valve seat/guide alignment is exact. See paragraph 7. Note that each guide has a locating clip fitted to it. This need not be removed before drifting out the worn guide but should be transferred to the new guide before it is drifted into the cylinder head.

5   Check that each valve stem is not bent by placing it against a straight-edge. If found to be bent the valve must be renewed. This check must be carried out if it is suspected that the engine has been over-revved.

6   The valves must be ground to provide gas-tight seal, during normal overhaul, or after recutting the seat or renewing the valve. Valve grinding is a simple, but laborious task. Smear grinding paste on the valve seat and attach a suction grinding tool to the valve head. Oil the valve stem. Rotate the valve in both directions, lifting it occasionally and turning it through 90°. Start with coarse paste, if the seats are badly pitted, and continue with fine paste until there is an unbroken grey ring on each seat and valve. Wipe off very carefully all traces of grinding paste. If any remains in the engine, it will cause very rapid wear!

7   If after regrinding, the valve seats become pocketed or if the valve seat width exceeds 1.4 mm regrinding will be necessary

to restore the seat condition. As shown in the accompanying illustration, three cutters or stones of different cutting angles are required to correct the seat profile and width. Because of the cost of these items and the skill their successful use requires, it is recommended that the cylinder head be returned to a competent motorcycle engineer who can carry out the work. Seat cutting will also be required after new valve guides have been fitted.

8   Before reassembling the valve components, inspect each component for signs of wear or failure. Measure the valve spring free length and renew any spring that has set to a shorter length than its service limit, bearing in mind that the inner and outer springs should be renewed as a pair.

Inner valve spring free length	40.2 mm (1.5827 in)
Service limit	39.0 mm (1.5354 in)
Outer valve spring free length	43.75 mm (1.7225 in)
Service limit	42.5 mm (1.6732 in)

If in any doubt as to the condition of a component, consult an official Honda Service Agent who will be able to provide advice and obtain a new item for comparison.

9   Check the cylinder heads for warpage by placing them on a surface plate or plate glass sheet and measuring the amount of warpage with a feeler gauge. If wapage exceeds the service limit the cylinder head will have to be ground or renewed.

Cylinder head maximum warpage	0.1 mm (0.0039 in)

Cylinder head/cylinder block mating surface warpage is usually due to uneven tightening of the cylinder head bolts.

10 To reassemble the valve components, fit new oil seals to each valve guide after checking that the valve spring seats are in place. Oil the valve stem, and fit the valve. Fit both valve springs with the close coils next to the cylinder head, followed by the valve spring collar. Compress the springs and fit the valve collets. Relax the springs, ensuring that the collets remain in position. Check that the collets are correctly located and seat them by sharply tapping the end of the valve stem with a hammer. The cylinder head should **not** be resting flat on the bench when doing this.

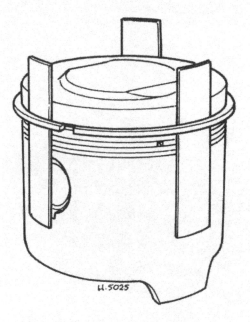

H.5025

Fig. 1.7 Freeing gummed piston rings

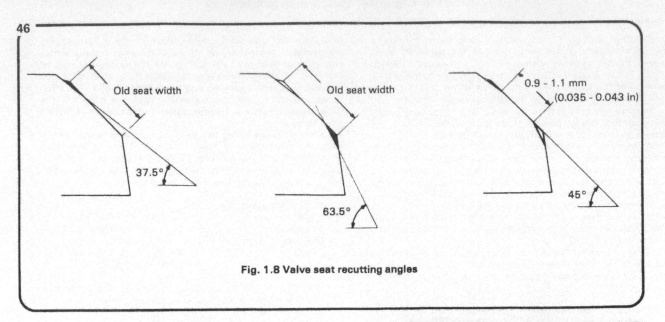

Fig. 1.8 Valve seat recutting angles

28.1a Remove the rocker arm/spindle support ...

28.1b ... the end cap (left-hand cylinder head only) ...

28.1c ... and the camshaft together with the oil seal

28.9a Fit the valve spring seat, followed by a new oil seal ...

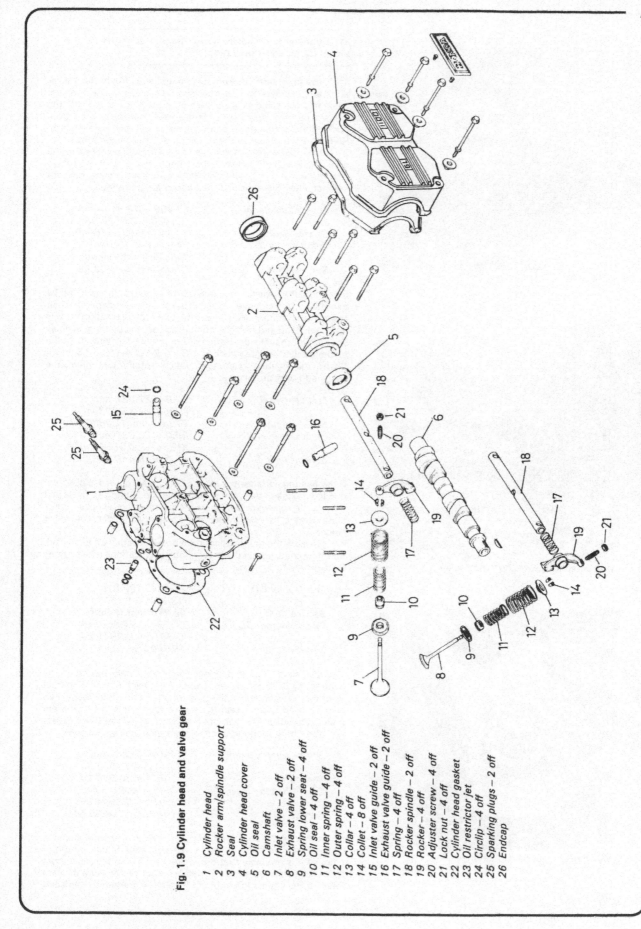

Fig. 1.9 Cylinder head and valve gear

1 Cylinder head
2 Rocker arm/spindle support
3 Seal
4 Cylinder head cover
5 Oil seal
6 Camshaft
7 Inlet valve – 2 off
8 Exhaust valve – 2 off
9 Spring lower seat – 4 off
10 Oil seal – 4 off
11 Inner spring – 4 off
12 Outer spring – 4 off
13 Collar – 4 off
14 Collet – 8 off
15 Inlet valve guide – 2 off
16 Exhaust valve guide – 2 off
17 Spring – 4 off
18 Rocker spindle – 2 off
19 Rocker – 4 off
20 Adjuster screw – 4 off
21 Lock nut – 4 off
22 Cylinder head gasket
23 Oil restrictor jet
24 Circlip – 4 off
25 Sparking plugs – 2 off
26 Endcap

28.9b ... insert the valve ...

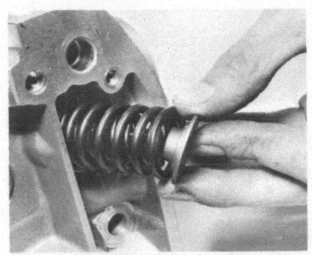

28.9c ... fit the valve springs with the spring collar ...

28.9d ... and retain them in position with the collets

29 Camshafts, rocker arms and rocker spindles : examination and renovation

1   Dismantle both rocker spindle holders. Mark the rocker arms and spindles so that they can be refitted in their original positions. The spindles are a push fit in the holders and with the holders removed from the cylinder head, can be tapped from position from the front edge, using a suitable parallel drift. Remove each rocker arm and spring as they become free.

2   Check the outside diameter of each rocker spindle and the bore of the spindle holes in the spindle holder and the rocker arm. Excessive wear will necessitate renewal of the relevant components. The specified clearances are as follows:

Rocker spindle OD	13.966 – 13.984 mm (0.5498 – 0.5505 in)
Service limit	13.94 mm (0.5488 in)
Spindle hole ID	14.0 – 14.018 mm (0.5512 – 0.5519 in)
Service limit	14.05 mm (0.5532 in)

3   In order to check the camshaft bearing diameters it will be necessary to refit the spindle/rocker arm holders on their respective cylinder heads. Insert the two hollow locating dowels into each head and refit the holders so that the bearing oil holes are on the exhaust side. Refit and tighten the holder bolts to the recommended torque which is 1.8 – 2.2 kgf m (13 – 16 lbf ft). Check the bearing diameters which should be within the specified limits as follows:

End bearing ID	27.0 – 27.021 mm (1.0630 – 1.0638 in)
Service limit	27.050 mm (1.0650 in)
Centre bearing ID	25.0 – 25.021 mm (0.9843 – 0.9851 in)
Service limit	25.050 mm (0.9862 in)

If the bearing surfaces are scored or appear to have 'picked-up' due to lubrication failure, it is likely that the bearing surfaces will have to be renewed. Replacement bearings are supplied complete with a cylinder head, rocker spindle holder and valve guides.

4   Measure the journal diameters on each camshaft and check that they come within the prescribed specifications which are as follows:

End journal OD	26.954 – 26.970 mm (1.0612 – 1.0618 in)
Service limit	26.910 mm (1.0594 in)
Centre journal OD	24.934 – 24.950 mm (0.9817 – 0.9823 in)
Service limit	25.910 mm (0.9807 in)

The camshaft journal/bearing clearances can be found by subtracting the measurement taken from each camshaft journal from that of each respective bearing diameter. 'Plastigage' can also be used to ascertain the clearance. In this case separate measurement of the journals and bearings will not be necessary. The correct journal/bearing clearances are as follows:

End bearing clearance	0.030 – 0.067 mm (0.0011 – 0.0026 in)
Service limit	0.140 mm (0.055 in)
Centre bearing clearance	0.050 – 0.087 mm (0.0020 – 0.0034 in)
Service limit	0.140 mm (0.0055 in)

5   Inspect cam lobes for scoring or uneven wear and chipping of the hardened outer surfaces. If a cam lobe is badly worn or chipped it is probable that the rocker arm foot that runs on that lobe is also damaged and will require renewal. Measure the cam lobes at the highest point of lift (maximum diameter). Camshaft

renewal will be necessary if one or more lobes does not come within the specified limits.

Inlet lobe service limit	36.9 mm (1.45 in)
Exhaust lobe service limit	36.6 mm (1.44 in)

6 Refitting the rocker arms and spindles in the holders is a reversal of the dismantling procedure. Before refitting the camshafts in the cylinder heads, lightly lubricate each bearing surface with clean engine oil. Note that the camshaft fitted with the tachometer drive gear must be refitted in the right-hand cylinder head. Lubricate each camshaft seal with engine oil and fit one to the forward end of each camshaft, spring side inwards.

7 To guarantee an oil tight seal between the rocker holder and cylinder head, a small amount of jointing compound should be spread on each mating surface before fitting. Fit each rocker holder over its camshaft and fit the six retaining bolts, tightening them evenly in a diagonal sequence, starting from either of the two most central bolts. Whilst doing this, ensure that the oil seal is pushed inwards as far as possible. Torque load the bolts to a setting of 1.8 – 2.2 kgf m (13 – 16 lbf ft) and wipe any excess jointing compound from around the joint. Finally, fit the sealing cap into its location in the rear of the left-hand cylinder head and fully loosen each valve adjuster screw in preparation for fitting the cylinder head to the cylinder block.

29.6 Note the camshaft marking denoting left or right-hand

30 Oil seals and O-rings : renewal

1 It is recommended that all oil seals and O-rings be renewed whenever the engine is dismantled. This is particularly important with those seals and rings that are non-accessible where major dismantling would be required for subsequent renewal. Seals which have previously given faultless service often begin to weep after reassembly due to damage during handling. This is particularly so with seals which are removed or refitted over splined shafts.

2 In most cases the oil seals are fitted between two separate cases, and can be removed easily after separation of the case. Other seals are a light drive fit in their housings and can be prised out of position using the flat of a screwdriver. Removal in this manner will invariably render the seal useless.

31 Cam belt pulleys and cambelts : examination and renovation

1 Wear of the cam belt pulleys is minimal and they will not require renewal until a considerable mileage has been covered. Check the profile of the teeth against that of a new belt. If

excessive wear is evident, the pulley wheel should be renewed without question as a worn pulley will accelerate wear of the cam belt and cause inconsistencies in the valve timing.

2 The toothed camshaft belts consist of fibreglass cores binding a synthetic rubber cover. Check wear of the teeth, renewing if there is any doubt about their condition. The belts are immensely strong under extension forces but are easily damaged if mishandled. Do not bend any part of the belt to a radius of less than approximately 25.0 mm (1.0 in) and never try and bend the belt about the centre line. Minor scores or scratches caused by removal using screwdrivers or other levers will usually develop into severe damage causing early failure of the belt. Again, renew if there is any cause for doubt.

3 The belt jockey wheel tensioners will rarely give trouble and only require renewal if the outer periphery becomes damaged, thereby causing wear of the pulley belts, or if the centre bearings fail. In either case, the complete wheel and bracket must be renewed. The extension springs fitted to the pulleys automatically control the tension of each belt when they are adjusted manually. If it is evident that the springs have become weak and are not giving sufficient load, they should be renewed.

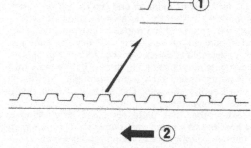

Fig. 1.10 Examination of cambelt
1   Damaged tooth   2   Direction of belt travel

32 Primary drive chain : examination and renewal

1 Examine the primary drive chain for wear and loose or broken side plates. The chain is of the Hy-vo type and therefore does not have rollers. There are no specific figures available by which wear can be assessed, but some indication of wear can be obtained by the amount of wear in the chain guides. If the chain guides are very badly worn it is evident that the chain has stretched. Renew both guides and the chain.

33 Clutch assembly : examination and renovation

1 Check the teeth on the duplex sprocket to the rear of the clutch outer drum for chipping or wear. As with the other chain drives in this engine, the components work under almost ideal conditions and will therefore have a very long service life.

2 Carefully clean all the clutch plates. Check the plain plate warpage by placing each plate on a flat surface and measuring with a feeler gauge. If the plates show signs of 'blueing' or bad scoring they should be renewed.

Maximum plate warpage	0.15 mm (0.006 in)

3 Measure the thickness of each friction plate with a vernier gauge. The correct dimensions are as follows:

Standard thickness	3.42 – 3.58 mm (0.1347 – 0.1410 in)
Service limit	3.20 mm (0.1260 in)

It is probable that all the friction plates will wear at a similar rate and should therefore be renewed as a complete set.

4 Check that the plate tongues that locate in the outer drum

and clutch centre are not worn; any serious burrs or indentations mean renewal. Very small burrs can be removed with a stone or a fine cut file. **Note:** Do not remove too much metal since the tongues will then be of unequal width and spacings, thus they will not take up the drive evenly and will wear the aluminium alloy clutch housing and centre.

5   Examine the clutch outer drum and centre boss grooves, removing any small burrs with a blind edge file. Burrs left in the two components will prevent the clutch from disengaging smoothly, causing noisy gear selection.

6   Measure the free length of each clutch spring. After prolonged service, the springs will take a permanent set (shorten) and will therefore exert less pressure.

Spring free length	*35.5 mm (1.3976 in)*
Service limit	*34.2 mm (1.3386 in)*

7   The clutch operating mechanism, which is contained in the clutch outer cover, will give little trouble if greased regularly. If wear is suspected, the mechanism may be removed from the cover by removing the split-pin, withdrawing the operating shaft and removing the adjusting arm. Inspect the cam on the operating shaft for wear or damage and renew it if necessary. Inspect the cup on the end of the clutch adjusting bolt for wear and cracking and renew if damaged. Although the short operating rod should suffer very little or no wear during its service life, it is advisable to check the ends for signs of damage or wear and along its length for straightness. Before reassembling the components, clean each item thoroughly and lightly grease the pivot bearing surfaces. The split-pin should always be replaced with a new item.

8   Finally, inspect the lifter plate for signs of cracking or damage. It is possible to impose very heavy loads on the plate by failing to remove or refit its six retaining bolts evenly and in a diagonal pattern; these loads may well break the plate. Push the operating rod guide from the centre of the bearing and inspect it for wear or damage. The bearing should be cleaned thoroughly in petrol and checked for wear. If roughness is felt when it is rotated the bearing should be renewed. Removal of the bearing from the lifter plate is easily achieved by supporting the plate close to the bearing edge and drifting the bearing out of the plate with a length of tube or a socket of the same diameter as that of the bearing inner race. Fit the new bearing by using a similar method to that for removal. Use a length of tube or a socket of the same diameter as that of the bearing outer race and take care to ensure that the bearing enters the plate squarely. Lubricate the bearing thoroughly in engine oil

before pushing the operating rod guide back into position in its centre.

34 Gearbox components : examination and renovation

1   Examine the gearbox components cery carefully, looking for signs of wear, chipped or broken teeth and worn dogs or splines.

2   If a gear requires renewal, it is probable that the gear with which it meshes will also be worn or damaged in a similar manner. Both gears should be renewed at the same time to prevent problems due to uneven wear between a new component and a partially worn one. Removal of the gear pinions from the mainshaft can only take place after the front bearing has been withdrawn. This can be done using a two-or three-legged sprocket puller positioned so that the 33 tooth top gear pinion is drawn off the shaft at the same time. The remaining gear pinions can then be removed in the same way as those on the layshaft, by removal of the various splined thrust washers and circlips. Carefully note the sequence of gears, washers and circlips for reference when reassembling. Lay each component on a clean worksurface, alongside the component removed previously. Correct positioning of each component is essential during reassembly; if in any doubt as to the location of a component, refer to the figure accompanying this text.

3   If the fit on any of the gears on their shafts is suspect, measure the relative shaft diameters and gear pinion internal diameters and compare them with the following specifications. Renew where necessary.

Shaft to mainshaft 4th gear and layshaft 2nd and 3rd gear pinion clearance	*0.040 – 0.082 mm (0.0016 – 0.0032 in)*
Service limit	*0.182 mm (0.0071 in)*
Thrust bush to mainshaft 4th gear and layshaft 1st gear clearance	*0.040 – 0.082 mm (0.0016 – 0.0032 in)*
Service limit	*0.182 mm (0.0071 in)*

4   Clean the journal ball bearings thoroughly in petrol and check for wear. Renew bearings if roughness is felt when they are rotated or if pitting or up and down play (radial) is evident.

5   Roll the selector fork rod on a piece of plate glass to check

33.7 The clutch operating mechanism is contained within the clutch cover

33.8 Inspect the clutch lifter plate assembly

for straightness. If this is satisfactory, measure the outside diameter of the rod and the inside diameter of the selector fork bores. Also check that the finger thickness of the forks is within the recommended limits as follows:

Selector fork ID	13.0 – 13.027 mm (0.5118 – 0.5129 in)
Service limit	13.050 mm (0.5138 in)
Selector rod OD	12.966 – 12.984 mm (0.5105 – 0.5112 in)
Service limit	12.90 mm (0.5079 in)
Fork finger thickness	6.4 – 6.5 mm (0.2520 – 0.2559 in)
Service limit	6.1 mm (0.2402 in)

6 If the machine tends to jump out of gear it is most probably due to worn dogs on the gear pinions. Difficulty in selecting gears is usually caused by bent selector forks or worn fork guide channels in the change drum. Check that the change drum channels are not worn beyond the specified limits.

Standard groove width	13.0 – 13.018 mm (0.5118 – 0.5125 in)
Service limit	13.04 m (0.5134 in)

Visual indication of wear in the change drum channels is usually most evident at abrupt changes of curvature.

7 Wear in the gear selector mechanism can only be rectified by direct replacement of the parts concerned. This applies equally to those components on the outside of the gearbox, such as the stop arms. If jumping out of gear or over selection has been experienced, renew the stop arm and change the claw springs as a first step in eliminating the causes. This can be done while the engine is still in the frame. If the pins in the end of the change drum become worn they too can be renewed after removal of the end cap. The cap is held by a counter-sunk screw. Note that one pin is of double diameter and of a shorter length than the other four. Breakage of the main change arm centraliser spring can only be remedied after complete dismantling of the engine. For this reason this spring should be renewed if there is the least doubt as to its condition.

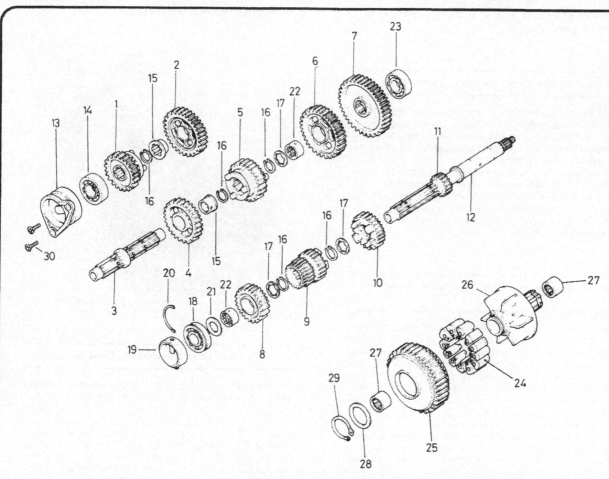

Fig. 1.11 Gearbox components and primary shaft

1	Layshaft 5th gear pinion	11	Mainshaft 1st gear pinion	21	Thrust washer
2	Layshaft 2nd gear pinion	12	Mainshaft	22	Splined collar
3	Layshaft	13	Bearing holder	23	Layshaft bearing
4	Layshaft 3rd gear pinion	14	Layshaft bearing	24	Damper blocks
5	Layshaft 4th gear pinion	15	Bush – 2 off	25	Primary driven gear
6	Layshaft 1st gear pinion	16	Circlip	26	Gear hub
7	Final drive pinion	17	Splined washer	27	Needle roller bearing – 2 off
8	Mainshaft 5th gear pinion	18	Mainshaft bearing	28	Thrust washer
9	Mainshaft 2nd and 3rd gear pinion	19	Mainshaft oil seal	29	Circlip
10	Mainshaft 4th gear pinion	20	Bearing half-ring	30	Screw – 2 off

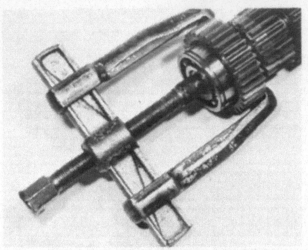

34.2 A puller must be used to remove the mainshaft front bearing and top gear pinion

36.2a Remove the circlip and thrustwasher ...

35 Primary drive gears : examination and renovation

1   Examine the condition of the gears upon which the 'Hy-vo' chain runs. The chain gear operated under almost ideal conditions in that lubrication is supplied in adequate quantities, and the components do not suffer from condensation as is often the case with primary drive systems.

2   Wear in either of the two gears will necessitate renewal. Unfortunately, if the drive gear requires replacement, the crankshaft will also have to be renewed as the two components are considered as a single unit and are not supplied separately. Removal of the driven gear housing from the hub assembly is described in the following Section.

36 Primary driven gear assembly : examination and renovation

1   The primary driven gear has contained within it a damper mechanism which consists of rubber blocks acting as buffers between fins cast in both the gear hub and gear housing. Any surge or roughness in the transmission may well indicate that the rubber blocks are worn; this may be confirmed by attempting to rotate the housing and hub in opposite directions, any excessive movement between the two will indicate that the blocks must be taken out and renewed.

2   To gain access to the rubber blocks, remove the circlip from the housing end of the assembly, followed by the thrust washer located beneath it. The hub may then be drawn out of the housing from the opposite end, leaving the rubber blocks contained within the housing. Replace the blocks with new items, noting that each block has two projections on one of its ends; these projections must face the gear housing side.

3   Reassembly of the housing is a reversal of the dismantling procedure. It will be found that canting the upper ends of the blocks away from each other, as described in Section 39 for the alternator shaft, will make the relocation of the hub into housing less difficult. Ensure the circlip is correctly located in the hub groove.

4   Check the two needle roller bearings in the hub centre for wear. If worn, each bearing may be drifted out from the opposite end of the hub. The journal ball bearing should also be checked for wear and if necessary, drawn from position on the mainshaft by using a two-legged sprocket puller.

36.2b ... to allow access to the rubber blocks contained within the unit housing

37 Final drive output shaft : examination and renovation

1   Check the shaft double gear for worn, broken or chipped teeth. Renew if necessary. Note that the smaller of the two gears is in fact a splined boss which engages with the moving portion of the shock absorber unit.

2   The shock absorber system comprises two cam-faced bosses loaded by a helical spring, which is retained under tension by an end cap and two split cotters. Removal of the spring requires the use of a special tool by which the spring may be compressed and the retaining cotters displaced. The system of spring retention is also used on most types of rear suspension unit. If the special tool is not available, the type of clamp used for suspension unit dismantling could probably be used. Alternatively, a pair of scarf joint clamps as used by carpenter joiners can also be utilised.

3   Compress the main spring sufficiently to release the retaining cotters and then release the tension slowly. Do not overcompress the spring as it may be damaged permanently. Examine the cam faces for excessive indentation or flaking. Although alteration of the cam profile due to wear will have little effect on the performance of the shock absorber it is probably wise to renew the two mating components if wear has

progressed to a point when the case hardening has been worn through.

4   With the two cam components drawn off the shaft, their splined inner faces may be inspected for wear along with the splined portion of the shaft. If the splines show signs of excessive wear or are chipped or damaged in any way, thus preventing easy movement of the two cam components along their length, then the shaft assembly should be returned to an official Honda Service Agent for inspection and renewal.

5   Measure the free length of the helical spring, renewing if it fails to meet the specification for length.

Spring free length   110.9 mm (4.3661 in)
Service limit   110.0 mm (3.9370 in)

6   Reassembly of the shaft components is a reversal of the dismantling sequence. Lightly oil the splined portion of the shaft before fitting the cam components. Note that the ball bearings for the shaft will be retained in the crankcase and rear engine cover, and should be inspected as described in Section 40 of this Chapter.

38.1 Inspect the starter clutch springs and end caps

38 Starter clutch : examination and renovation

1   The starter clutch assembly is housed in the rear of the alternator generator rotor. Check the condition of the three engagement rollers, spring cap and springs. These components may be dislodged from their housings by carefully hooking out the rollers with the flat of a small screwdriver; this will allow the cap to be pushed forward by the expanding spring thus allowing them both to be inspected in situ. If the springs show signs of fatigue or are broken then they must be renewed. It is unlikely that either the rollers or caps will require attention as they are only subjected to a limited amount of use. On completion of inspection, carefully push each cap back into its housing, so that the spring is fully compressed, and insert each roller in turn. If necessary, the complete starter clutch can be removed from the rear of the rotor after unscrewing the three countersunk screws.

2   Examine the starter drive sprocket and driven sprocket for worn or damaged teeth and the drive chain for wear. In the unlikely event of wear having taken place, renewal of the components is the only method of renovation, bearing in mind that under no circumstances should a worn component be refitted and run with a new component.

39 Alternator damper unit : examination and renovation

1   The alternator rotor and starter clutch are mounted on a double gear shaft incorporating a rubber segment cush drive shock absorber system. The gear pinions and cush drive unit are retained by a short helical spring in a manner similar to that used on the final drive output shaft.

2   With the aluminium bearing housing removed from the shaft, the unit may be dismantled by compressing the spring and releasing the two collets from the spring retaining plate. Some force will be required to compress the spring and this should not be attempted unless a tool similar to the one shown in the accompanying photograph is available. Serious personal injury may result if the spring is not properly secured during collet removal.

3   Remove the circlip securing the damper plate to the shaft and withdraw the damper plate followed by the gear pinion, stopper plate and second gear pinion. Examine the condition of the two gears, the damper plate and stopper plate, renewing if required. Check the cush drive rubbers for hardening or compac-

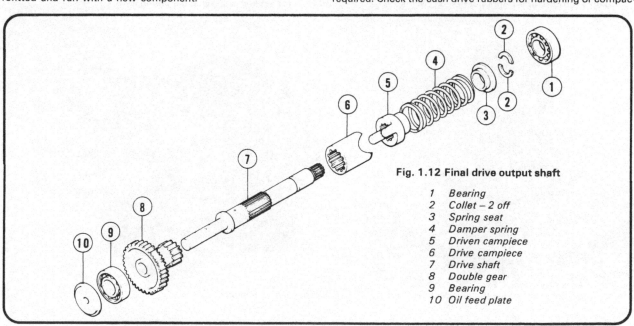

Fig. 1.12 Final drive output shaft

1   Bearing
2   Collet – 2 off
3   Spring seat
4   Damper spring
5   Driven campiece
6   Drive campiece
7   Drive shaft
8   Double gear
9   Bearing
10 Oil feed plate

tion and renew if either is evident. Measure the free length of the helical spring and renew if it has taken a permanent set.

Spring free length   26.0 mm (1.023 in)

Before reassembly, check the journal ball bearing for wear. Renew the bearing if any up and down (radial) play or roughness and pitting is evident.

4   Reassembly of the unit is a reversal of the dismantling sequence. Note the positioning of the rubbers in the gear pinions as shown in the accompanying photograph; fitting of the damper plates to the gear pinions will be almost impossible unless the rubbers are positioned in this way. When fitting the stopper plate, ensure the pins protruding from the plate locate correctly in the pinion holes. Note also the thrustwasher fitted beneath the retaining circlip.

39.2a Use a spring compressor tool ...

39.2b ... to allow removal of the spring retaining collets ...

39.2c ... the retaining ring and the spring itself

39.3a Remove the circlip to release the damper plate

39.3b ... the gear pinion ...

39.3c ... the stopper plate and the second gear pinion

39.4 Position the rubbers in the gear pinion as shown

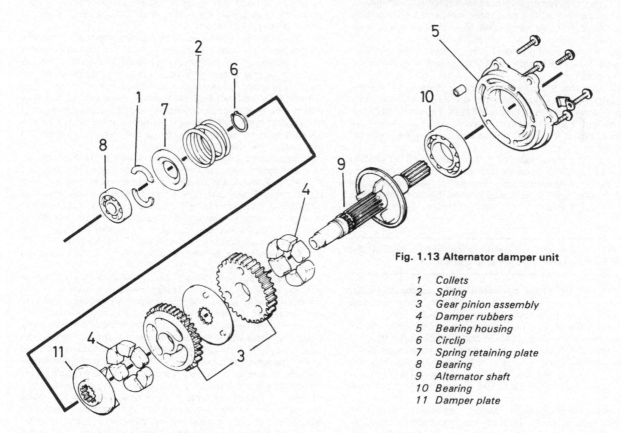

Fig. 1.13 Alternator damper unit

1 Collets
2 Spring
3 Gear pinion assembly
4 Damper rubbers
5 Bearing housing
6 Circlip
7 Spring retaining plate
8 Bearing
9 Alternator shaft
10 Bearing
11 Damper plate

40 Journal ball bearings : examination, removal and fitting

1 When the engine is in a completely dismantled state check the condition of all journal ball bearings.

2 Wash each bearing thoroughly with petrol so that all old oil and any foreign matter has been removed. Allow the bearings to dry. Test the bearing for roughness by spinning the outer race. Any roughness or snatching indicates wear or pitting in the races. The bearing should therefore be renewed. Bearings which are an interference fit in their housings should be tested in situ as the outer races are designed to compress slightly when correctly positioned, giving the correct original tolerances. There should be no radial clearance on ball bearings though a small amount of side play is acceptable and on some bearings is evident even when new.

3 Bearings fitted to shafts may be removed, using a two- or three-legged sprocket puller. In some cases the clearance behind the bearing is not sufficient to allow purchase of the puller legs. In these cases a special ball bearing puller tool should be used.

4 Removal of bearings within a housing in the crankcase or in detachable housings should only be made after heating the case – using a blowtorch or by placing the casing in question in an oven. The correct temperature is 100° – 150°C. If a blowtorch is used care should be taken not to overheat the case locally as

this may damage the alloy or cause permanent distortion. Most bearings can be tapped from position, using a suitable drift. The use of socket spanners for this purpose is invariably decried and equally invariably used. Bearings which are located in blind housings, as is the case with the gearbox layshaft front bearing, can be removed after heating the case to the correct temperature and tapping the casing on a block of wood. The bearing will then fall free. New bearings may be tapped into place after heating the casing as for removal.

41 Engine/gearbox unit reassembly : general

1 Before reassembly of the engine/gear unit is commenced, the various component parts should be cleaned thoroughly and placed on a sheet of clean paper, close to the working area.

2 Make sure all traces of old gaskets have been removed and that the mating surfaces are clean and undamaged. One of the best ways to remove old gasket cement is to apply a rag soaked in methylated spirit or where necessary, a gasket cement solvent. This softens the cement allowing it to be removed without resort to scraping and consequent risk of damage.

3 Gather together all the necessary tools and have available an oil can filled with clean engine oil. Make sure all new gaskets and oil seals are to hand, also all replacement parts required. Nothing is more frustrating than having to stop in the middle of a reassembly sequence because a vital gasket or replacement has been overlooked.

4 Make sure that the reassembly area is clean and that there is adequate working space. Refer to the torque and clearance settings whenever they are given. Many of the smaller bolts are easily sheared if over-tightened. Always use the correct size screwdriver bit for the crosshead screws and never an ordinary screwdriver or punch.

5 Ensure also that a tube of silicone rubber (RTV) jointing compound is available because this is used in place of gaskets, particularly in the case of the crankcase joint. Also required is a tube of thread locking fluid which must be used to secure certain screws in position.

42 Reassembling the engine/gearbox unit: refitting the gearbox components, pistons and crankshaft

1 Position the right-hand crankcase on the workbench so that it is resting on the cylinder block mating surface and ensure that the primary drive chain guide is correctly fitted.

2 Lubricate the change drum bearing surfaces and introduce it into position through the hole in the gearbox wall. Insert the neutral indicator switch, with its O-ring, into its location in the crankcase and secure it in position with the small retaining plate and single retaining bolt. The drum is now retained in the casing by the switch and should be turned so that it is in the neutral position.

3 Lubricate the layshaft bearing contained within the crankcase and insert the layshaft into the gearbox complete with all the gear pinions except the forwardmost one (top gear, 30T). If the layshaft assembly was dismantled for examination or renewal of gear pinions, it must be reassembled before refitting it in the gearbox. Refer to the accompanying illustration for the relative positions of the gears, washers and circlips. It is very important that the oil hole in each splined bush aligns with the corresponding oil hole in the shaft. With the layshaft in position refit the top gear pinion (30T) so that the dogs face inwards. Lubricate the outer journal ball bearing with clean engine oil. Position the bearing, complete with bearing holder, so that the two screw lugs align with the holes in the gearbox wall. Carefully push the bearing holder home. Ensure that the threads of the two countersunk retaining screws are clean and coated with locking fluid before inserting and tightening the screws.

4 Insert the selector fork rod through the gearbox wall, fitting the three selector forks as it is pushed home. It is important that the three selector forks are fitted in the correct order on the shaft and the correct way up. Refer to photo 42.11b for indication. When the selector fork rod is fully home and the two outer forks are correctly engaged with the pinions on the layshaft, rotate the rod by means of the screwdriver slot in the outer end so that the locating pin holes align. Insert the pin through the crankcase wall and gently tap it home until the head of the pin is just lower than the edge of the crankcase mating surface. Do not knock the pin in further than necessary or subsequent removal will be made difficult.

5 Reposition the crankcase so that access can be made to the right-hand cylinder bores. Refit the piston rings onto the right-hand pistons. When refitting the 'sandwich' type oil scraper ring, the two thin plain rings must be positioned so that their gaps are 20 mm (0.80 in) or more from the gap of the corrugated ring and more than 40 mm (1.60 in) from each other's gaps. The upper rings must be fitted so that the letter mark, which is stamped on one side of each ring, faces upwards. Also, the profiles of the top rings differ so be sure to replace them in the correct groove. This is important to retain maximum compression. When fitting any piston into its cylinder bore, the rings should be arranged so that the end gaps are approximately 120° apart.

6 Lubricate the cylinder bores with clean engine oil. Insert the pistons complete with connecting rods and rings into cylinders No 1 and No 3. A piston ring clamp should be used to compress the rings as they enter the bores. Fitting the pistons without a clamp is possible but the risk of damage to the rings is great as there is no chamfered lead-in. The pistons should be fitted so that the oil hole in each connecting rod faces the top edge of the engine. Note that although oil holes are provided on the connecting rods for No 1 and No 3 cylinder, no oil holes are provided in the shell bearings for these rods.

7 Reposition the crankcase so that access to the gearbox can be made again. Lubricate the big-end journals on the crankshaft and fit No 2 and 4 pistons and connecting rods complete with piston rings. Ensure that the big-end bearing shells are fitted correctly so that the oil holes in the shells align with those in the connecting rods. Fit the connecting rods to the crankshaft so that the oil holes face the top edge of the engine. Tighten the bearing cap nuts evenly to a torque of 3.0 – 3.4 kgf m (22 – 25 lbf ft), checking as tightening progresses that the bearings remain free.

8 Place the big-end shells of No 1 and 3 connecting rods in position in the rods and the bearing caps. In the same way fit the main bearing shells into their respective bearing caps. It is absolutely essential that all surfaces of the bearing shells on any part of the engine and the caps into which they fit are perfectly clean. Lubricate all the exposed journals on the crankshaft with clean engine oil. Lubricate and fit the oil seal onto the front end of the crankshaft and position the 'Hy-vo' primary chain on the primary drive sprocket (gear). If a part-worn chain is being used ensure that it is refitted so that it runs in the original direction of rotation. If this requirement is not observed excessive noise and accelerated wear will result. Lift the complete crankshaft assembly and place it carefully into position in the crankcase.

9 Insert the main bearing cap hollow dowels and fit the main bearing caps. Note the marks made on the caps during the dismantling procedure and ensure that each cap is fitted in its original position. The arrow marked on each cap must face the engine top edge. Position the big-end bearings of No 1 and No 3 cylinder on the journals and refit the bearing caps. Tighten the cap retaining bolts of the two end main bearings to a torque of 4.8 – 5.2 kgf m (35 – 38 lbf ft) and the cap retaining bolts of the centre main bearing to a torque of 6.7 – 7.3 kgf m (49 – 53 lbf ft). Tighten the big-end bearing cap retaining nut of No 1 and No 3 cylinders to a torque of 3.0 – 3.4 kgf m (22 – 25 lbf ft). Rock the crankshaft occasionally as tightening proceeds to ensure that the bearings do not tie.

10 If the gearbox mainshaft was dismantled it must now be reassembled, complete with bearings, before being fitted in the

casing. As with the layshaft assembly, fit the various gear pinions, washers and circlips by referring to the accompanying illustration. Again the splined bush is provided with an oil orifice which must align with the hole in the shaft. When the bearings have been refitted, check that the exact distance between the outer facing faces of the two journal ball bearings is 177.0 mm (6.969 in). This is to ensure that the bearing half clips in the crankcase align with the radial locating grooves in the bearing outer races. For the same reason, the bearings must be fitted the correct way round as the radial grooves are offset in the outer race width.

11 Insert the mainshaft bearing half clips in the grooves in the crankcase half cups. Lubricate the primary driven gear internal needle roller bearings with engine oil and mesh the gear with the Hy-vo chain. Insert the mainshaft through the primary driven gear and lower the complete assembly onto the crankcase so that the bearings locate with the half clips and the central selector fork locates correctly in the central groove of the 2nd/3rd gear pinion. Refit the mainshaft blind end cap so that the small rubber locator tab aligns with the recess in the gearbox wall. This is important as it ensures that the oil passage orifice in the end cap aligns with the feed channel in the gearbox wall. Finally, refit the oil supply nozzle in its location next to the primary driven gear.

42.2a Insert the change drum through the hole in the gearbox wall ...

42.2b ... and retain it in position with the neutral indicator switch

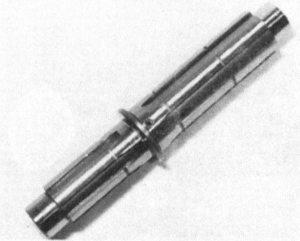

42.3a The gearbox layshaft

42.3b The oil hole in the splined bush must align with the hole in the layshaft ...

42.3c ... when the bush is fitted, together with the 3rd gear pinion

42.3d Retain the 3rd gear pinion with the circlip ...

42.3e ... and fit the 4th gear pinion ...

42.3f ... together with the circlip and splined thrustwasher

42.3g Align the oil hole in the splined bush with the hole in the layshaft ...

42.3h ... before fitting the bottom gear pinion ...

42.3i ... followed by the final drive gear pinion

42.3j Align the oil hole in the 2nd gear pinion bush with the hole in the layshaft ...

42.3k ... before retaining the bush and pinion with the circlip

42.3l Insert the layshaft assembly into the gearbox casing ...

42.3m ... fit the top gear pinion ...

42.3n ... followed by the bearing, complete with bearing holder ...

42.3o ... and the holder retaining screw

42.6a Fit the piston/connecting rod assembly ...

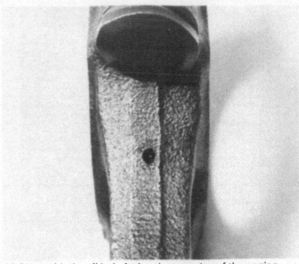

42.6b ... with the oil hole facing the top edge of the engine

42.7 Tighten the big-end bearing cap nuts to the correct torque loading

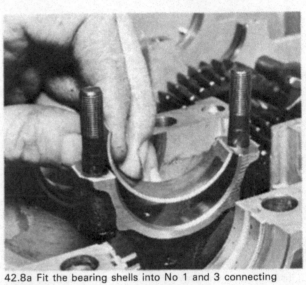

42.8a Fit the bearing shells into No 1 and 3 connecting rods ...

42.8b ... and into the main bearing cups ...

42.8c ... lubricating them with clean engine oil when fitted

42.9a Refit the big-end bearing caps to No 1 and No 3 connecting rods

42.9b Tighten the main bearing cap retaining bolts to the correct torque loading

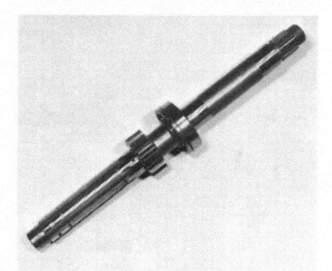

42.10a The gearbox mainshaft and 1st gear pinion with bearing

42.10b Note the condition of the bush within the 4th gear pinion ...

42.10c ... before fitting the pinion to the mainshaft ...

42.10d ... followed by the splined thrustwasher ...

42.10e ... and the retaining circlip

42.10f Fit the 2nd/3rd gear pinion ...

42.10g ... and retain it in position with the circlip followed by the splined thrustwasher

42.10h Ensure that the oil hole in the splined bush aligns with the hole in the mainshaft ...

42.10i ... before fitting the 5th gear pinion ...

42.10j ... followed by the thrustwasher ...

42.10k ... and mainshaft end bearing

42.10l Do not omit to fit the thrustwasher ...

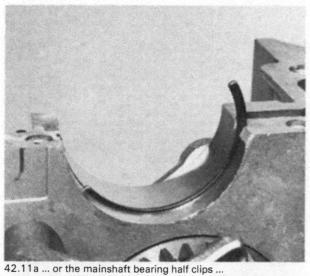

42.11a ... or the mainshaft bearing half clips ...

42.11b ... before fitting the complete mainshaft assembly into the gearbox casing

42.11c Note the position of the mainshaft blind end cap locator tab

42.11d Refit the oil supply nozzle

64

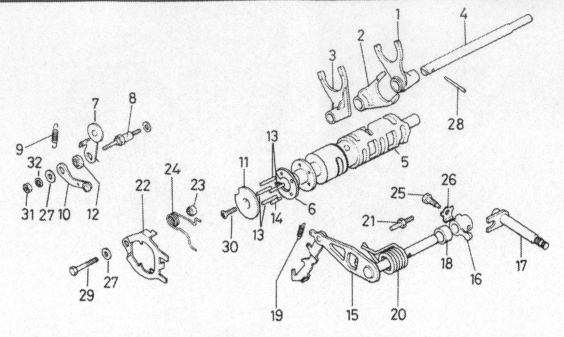

Fig. 1.14 Gearchange mechanism

1	Right-hand selector fork	9	Spring
2	Centre selector fork	10	Neutral stopper arm
3	Left-hand selector fork	11	Pin end-plate
4	Selector fork shaft	12	Centre boss
5	Gearchange drum	13	Change pin – 4 off
6	Roller stopper plate	14	Change pin
7	Stopper arm	15	Main change arm
8	Pivot stud	16	Ball-end link
17	Gearchange shaft	25	Locating bolt
18	Collar	26	Tab washer
19	Spring	27	Washer – 3 off
20	Centraliser spring	28	Pin
21	Spring anchor	29	Bolt
22	Main selector claw	30	Screw
23	Collar	31	Nut
24	Centraliser spring	32	Spring washer

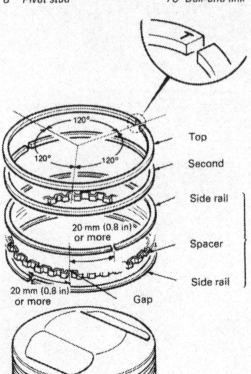

Fig. 1.15 Piston ring end gap positioning

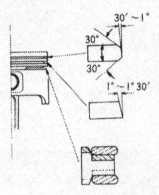

Fig. 1.16 Piston ring profiles

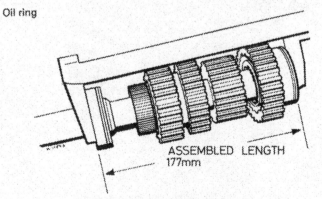

Fig. 1.17 Mainshaft assembled length

43 Reassembling the engine/gearbox unit : refitting the gearchange shaft

1   Position the left-hand crankcase on the workbench so that it is resting on the cylinder block mating surface. Insert the large oil catch plate into the crankcase and secure it in position with the lowermost retaining bolt, ensuring that the other two bolt holes in the plate are directly in line with the threaded holes in the crankcase., Refit the primary chain tensioner and secure it, and the oil catch plate, in position with the two retaining bolts.
2   Refit the small oil guide plate which is retained by two bolts passing through the crankcase to the rear of the flywheel viewing cap. Coat the two bolts with jointing compound before fitting them.
3   Grease the gearchange splined shaft and carefully push it through the crankcase wall from the inside. Do not rotate the shaft until it is fully home or the splines will damage the oil seal. Insert the gearchange main change arm into position in the front wall of the gearbox. Ensure that the centraliser spring distance piece is in place and that the two ears of the spring lie one either side of the anchor screw. As the change arm shaft is pushed fully home, fit the shaft ball end so that it engages with the fork on the splined gearchange shaft. Insert the retaining bolt, fitted with the tab washer, so that the bolt end locates with the radial hole in the shaft. Tighten the bolt and bend the tab washer upwards. Tie the main change pawl back against the change arm in a manner similar to that used for dismantling.
4   Move to the rearmost end of the crankcase and insert the oil scavenge pump rubber grommet into the hole adjacent to the mainshaft bearing half cap.

43.4 Insert the oil scavenge pump rubber grommet into the crankcase wall

44 Reassembling the engine/gearbox unit : joining the crankcase halves

1   Refit the three hollow crankcase locating dowels and the O-ring which is fitted to the smallest dowel. Generally lubricate all the internal engine components, including the cylinder bores, with engine oil.
2   Because the pistons on No 2 and 4 cylinders have to be fitted as the left-hand crankcase is lowered on the right-hand side and because access is obscured, no piston ring clamps can be used. To overcome this the base of each cylinder bore sleeve is heavily chamfered to aid refitting of the pistons.
3   It should be noted before commencing to join the crankcase halves, that the pistons must be supported so that they remain parallel and absolutely square as the upper crankcase half is

lowered into position. It was found that the most efficient method of ensuring that the pistons entered the bores cleanly was to have an assistant guiding each piston in turn into its bore after having positioned one piston higher than the other by rotating the crankshaft.
4   Coat the mating face of the left-hand crankcase half with jointing compound and lower the crankcase half down over the pistons so that the assistant can guide them into the cylinder bores. The casing may then be tapped home, using the palm of the hand. Fit the nineteen crankcase bolts that pass through the right-hand casing. There are three different sizes of bolt and their positions should have been noted when separating the crankcase halves by inserting them into a cardboard template. If this was omitted however, refer to the accompanying figure for the bolts' correct positions. Note when fitting these bolts, that the long bolt located to the rear of the left-hand cylinder barrels must have the special copper sealing washer located beneath its head. Tighten the bolts down evenly and in as much a diagonal sequence as possible to the following torque settings for each bolt size.

6 mm	1.0 – 1.4 kgf m (7 – 10 lbf ft)
8 mm	2.4 – 2.8 kgf m (17 – 20 lbf ft)
10 mm	3.3 – 3.7 kgf m (24 – 27 lbf ft)

Invert the crankcase and fit the remaining three crankcase bolts, torque tightening them to their correct settings, depending on size.

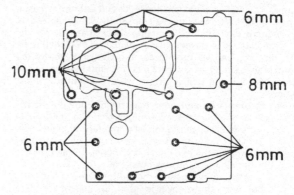

Fig. 1.18 Crankcase bolt positions

44.1 An O-ring is fitted to the smallest crankcase dowel

45 Reassembling the engine/gearbox unit : refitting the oil pumps

1   Insert the 8 mm and 14 mm hollow dowels in the main oil pump mounting lug and place a new paper gasket over them. Slide the oil pump driveshaft through the lug until the main pump is correctly positioned. Fit and tighten the pump retaining bolts.
2   Fit a new O-ring onto the inner boss of the clutch scavenge pump and position the pump over the shaft end. Push the pump home so that the oil outlet nozzle passes through the grommet in the casing wall. Fit and tighten the three pump retaining bolts. Check free movement of the pumps by rotating the shaft.

46 Reassembling the engine/gearbox unit : refitting the gear selector mechanism and front engine cover

1   Fit the main selector claw over the end of the change drum, with the tongues facing inwards. Make sure that the claw spring 'ears' are on the outside of the outer ears. Fit the centre boss so that the shoundered centre fits into the pivot hole in the claw. Refit the pivot bolt and plain washers. Assemble the two stopper arms, spring and washers in the correct sequence as shown in the illustration and fit the complete assembly onto the pivot stud so that the stopper rollers engage with the end of the change drum. Refit the washers and retaining nut. Untie the main change arm pawl and allow it to engage with the change pins.
2   Temporarily refit the gearchange lever and attempt to select each gear in turn. Rotation of the crankshaft will aid selection. If one or more gears will not engage, check that the external selector mechanism is correctly assembled. If this is found to be correct, the problem will be internal, requiring removal of the left-hand crankcase again. Suspect incorrect fitting of gears and spacer washers if the gear clusters were dismantled for examination.
3   Refit the one large and one small hollow dowel into the mating surface of the crankcase onto which fits the engine cover. Insert the feed collar into the oil pump face along with its O-ring. Fit the O-ring into the recess around the oil pump seal. Fit the O-rings around the two crankcase mating surface dowels and into the remaining crankcase recesses.
4   If the water pump has been removed for renewal, it should be refitted before the front cover is fitted. Ensure that the two O-rings on the pump body are correctly positioned before fitting the pump into the case. Fit and tighten the three retaining bolts. Place a new front cover gasket in position and refit the cover.

Rotate the water pump impeller as the front cover is being fitted, so that the pump shaft engages with the slot in the end of the oil pump shaft. Refit the cover screws in their previously noted positions, tightening them in an even and diagonal sequence.

47 Reassembling the engine/gearbox unit : refitting the alternator drive, starter clutch and alternator

1   With the engine unit supported on its base so that the rear face is tilted slightly upwards, insert the alternator drive shaft assembly through the crankcase wall and install four of the five retaining bolts. Omit the starter chain guide tab bolt.
2   It is necessary to adjust the backlash between the alternator shaft double gear and the crankshaft to prevent excessive noise when the engine is in service. Adjust the backlash to zero by hooking a spring balance around the alternator shaft and pulling in the direction of the crankshaft to a force of 1.0 ± 0.5 kg (2.20 ± 1.10 lb). Maintain this force and tighten the four housing retaining bolts to a torque of $1.0 - 1.4$ kgf m ($7 - 10$ lbf ft).
3   Lubricate the starter driven sprocket bearing with clean engine oil and refit the sprocket on the alternator drive shaft so that the boss faces outwards. Mesh the starter drive chain over the sprocket and refit the chain guide tab and retaining bolt. The splined thrustwasher should now be fitted over the splined end of the drive shaft so that it abuts against the shaft boss.
4   Lubricate the alternator bearings and refit the rotor onto the shaft so that the starter driven sprocket boss enters the starter clutch. Do not tap the starter clutch or alternator in an attempt to force the rollers over the boss. If the sprocket is rotated by pulling the chain whilst the rotor is held against it, the rotor will slide into place with ease. Fit the rotor retaining bolt and plain washer and tighten to a torque of $8.0 - 9.0$ kgf m ($58 - 65$ lbf ft) whilst preventing the rotor from turning by inserting a tommy bar in one of the balance holes drilled in the starter clutch housing. Mesh the starter motor drive sprocket with the starter drive chain.

48 Reassembling the engine/gearbox unit : refitting the final drive output shaft

1   Insert the output shaft double gear into the crankcase through the crankcase wall. Insert the final drive output shaft through the double gear, so that the shaft engages with the smaller gear (splined boss). Using a new gasket, refit the double gear access hole cover and fit and tighten the four retaining bolts

45.1 Insert the main oil pump and driveshaft into position

46.1 Fit the selector claw with the tongues facing inwards

46.3a Ensure all locating dowels ...

46.3b ... together with new O-rings are fitted before ...

46.4 ... refitting the engine front cover

47.3 Fit the splined thrustwasher so that it abuts against the shaft boss

47.4 Insert a tommy bar in the starter clutch housing balance hole to provide a means of preventing the rotor from turning

48.1 Insert the final drive output shaft to locate the final drive double gear

49 Reassembling the engine/gearbox unit : refitting the clutch assembly

1   Mesh the duplex oil pump drive chain over the sprocket on the clutch outer drum and fit the drum onto the splined clutch shaft. Before the outer drum is fully home, mesh the oil pump driven sprocket with the chain and fit the sprocket to the pump drive shaft. Refit and tighten the sprocket retaining Allen bolt and plain washer. Refit the 40 mm internal circlip into the centre of the outer drum, followed by the splined washer, which fits over the clutch shaft.

2   Place the clutch centre on the worksurface and fit the friction plate with the smaller width over it, followed by alternate plain and friction plates, finishing with a friction plate. Note that each plate should be lightly lubricated with clean engine oil and that the double thickness damper plate must be fitted between the 4th and 5th friction plates in place of a normal plain plate. To complete the assembly, fit the pressure plate so that its bosses protrude fully through the corresponding holes in the clutch centre.

3   The complete inner clutch assembly may now be fitted into the clutch outer drum by guiding the plate outer tongues into the slots of the drum whilst holding the assembly together by means of keeping a finger and thumb on one of the pressure plate bosses. Ensure that all the clutch plates are fully located in the outer drum and retain the inner assembly by fitting the special washer over the shaft end, followed by the slotted securing nut. The special washer must be fitted with the OUTSIDE mark outermost and the securing nut with the chamfered face inwards.

4   In order that the clutch shaft be prevented from rotating as the special nut is tightened, pressure should be applied to the plates using two or three clutch springs and bolts as described for dismantling. As before, a number of washers should take the place of the clutch lifter plate. Tighten the special nut to a torque setting of 5.5 – 6.5 kgf m (40 – 47 lbf ft).

5   Remove the springs and bolts which were refitted temporarily. Refit all the clutch springs followed by the lifter plate and the spring bolts, which should be tightened to a torque setting of 0.8 – 1.2 kgf m (6 – 9 lbf ft). Ensure the bolts are fitted in their previously noted positions and are tightened in small increments in a criss-cross pattern. Finally, lubricate and fit the short operating rod into the lifter plate guide.

50 Reassembling the engine/gearbox unit : refitting the rear engine cover and clutch cover

1   Refit the CDI pulser generator drive shaft into its location in the crankshaft end, ensuring that its locating spigot is pushed fully into the crankshaft and slot and that the O-ring is renewed and lightly lubricated before doing so.

2   Press the two hollow locating dowels into the crankcase mating face and carefully fit a new gasket over them. Grease the lips of the rear cover oil seals through which passes the engine final drive output shaft and CDI pulser generator drive shaft and refit the cover very carefully so that the ends of the shafts do not damage their respective seals. Refit the twelve cover securing bolts, not forgetting to position the electrical lead clamp and clutch cable guide in their previously noted positions, and tighten them evenly and in a diagonal sequence.

3   Fit a new gasket around the clutch assembly and press it firmly onto the rear cover mating surface. Place the clutch cover in position and fit and tighten the six securing bolts and two flange nuts, evenly and in a diagonal sequence.

51 Reassembling the engine/gearbox unit : refitting the CDI pulser generator assembly

1   Carefully refit the ATU locating pin into its recess in the crankshaft position. Push the thick insulator gasket into position

on the rear engine cover mating surface, taking care to guide it squarely over its locating pin. Follow this with the ATU assembly, ensuring that the slot in the base of the unit aligns with the pin in the crankshaft extension before pushing it into position. Fit the plain washer and retaining bolt and tighten the bolt.

2   Refit the pulser generator assembly and its housing and secure it in position with the three retaining bolts. Do not omit to fit the plain washers beneath two of the bolt heads. Finally, check that the electrical lead rubber grommet is correctly located in the housing wall and fit the housing cover complete with new gasket. Fit and tighten the three cover retaining bolts.

52 Reassembling the engine/gearbox unit : refitting the cylinder heads

1   Rotate the crankshaft by means of the alternator rotor bolt so that No 1 cylinder is at TDC on the compression stroke. Now rotate the crankshaft forwards or backwards through 90°. Check that each valve adjuster screw has been fully unscrewed. Carrying out these two simple precautions will serve to prevent the pistons coming into contact with the valves as the cylinder head units are fitted.

2   Each cylinder head unit, complete with valve gear, can be refitted using identical procedure. Commence by fitting either the right-hand or left-hand cylinder head, as chosen.

3   Ensure that the mating surfaces of the cylinder head and block are absolutely clean. Fit the two hollow locating dowels and the oil feed nozzle, which should be fitted with the shorter and narrower end towards the cylinder block. It is important that the O-rings are lubricated with clean engine oil, seen to be in good condition and are correctly fitted. Fit a new cylinder head gasket over the locating dowels and push it onto the cylinder block mating surface where it will adhere in position. Refit the cylinder head, followed by the six 10 mm retaining bolts. Ensure the two longer bolts are refitted in the positions noted during their removal. Tighten the six 10 mm bolts evenly, in a diagonal sequence, starting from either of the two most central bolts. Finally torque tighten the bolts down to 5.3 – 5.7 kgf m (38 – 41 lbf ft). Fit and tighten the remaining one 6 mm bolt to a torque of 1.0 – 1.4 kgf m (7 – 10 lbf ft).

4   Fit the crankcase to timing case inner extension seals and refit the extensions to the front of each cylinder head, fitting and tightening the two retaining bolts to secure each extension in position.

49.1a Refit the circlip (arrowed) into the centre of the clutch drum ...

49.1b ... followed by the splined washer

49.2a Fit the plates to the clutch centre, starting with the friction plate with the smaller width ...

49.2b ... fitting the double thickness damper plate between the 4th and 5th friction plates ...

49.2c ... and finishing with a friction plate

49.5 Fit the clutch lifter plate bolts in their previously noted positions (arrows indicate longer bolts)

50.1 Push the CDI pulser generator drive shaft into the crankcase end

50.2a Grease the lip of the final drive output shaft oil seal ...

50.2b ... before fitting the rear engine cover

51.1a Refit the locating pin into the crankshaft extension ...

51.1b ... followed by the insulator gasket and spark advancer assembly

51.2 Secure the pulser generator assembly in position with the three remaining bolts

52.3a Fit a new O-ring to the oil feed nozzle

52.3b With the new gasket fitted over the locating dowels ...

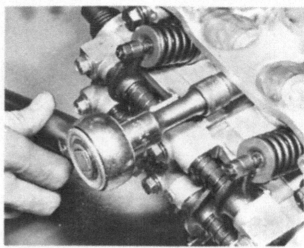

52.3c ... fit the cylinder head and torque load the retaining bolts

52.4a Fit the crankcase to timing case inner extension seals ...

52.4b ... and refit the extension to the front of each cylinder head

53 Reassembling the engine/gearbox unit : refitting the valve drive assembly and timing the valves

1   Refit the camshaft belt drive pulleys to the forward end of the crankshaft, ensuring that the Woodruff keys are correctly located and the plates and pulleys fitted in their correct order. Tighten the retaining bolt to a torque of 7.0 – 8.0 kgf m (51 – 58 lbf ft).
2   Refit the Woodruff key into the keyway in the end of each camshaft and fit the driven pulleys. The left-hand pulley should be with the boss facing towards the engine and the right-hand pulley boss should face outwards. When tightening the centre bolts, prevent the pulleys from rotating by placing a tyre lever through the pulley spokes so that it abuts against one of the timing case inner extension bolt heads.
3   Rotate the cam pulleys so that the arrow mark on the face of each is aligned with the raised timing marks on the corresponding case extensions. With a spanner on the alternator rotor bolt, rotate the crankshaft so that the 'T-1' mark on the crankshaft flywheel aligns with the index marks on the

observation hole. A length of wire can be placed across the aperture to aid accurate sighting. Without rotating the driven or drive pulleys fit the two timing belts. Old belts must be refitted in the same position as they were on dismantling. Do not strain the belts or prise them into position as damage may result, leading to short belt life. Fitting of the belts will be made easier by pulling the tensioner pulleys away from the belts and locking them in position by tightening the bolts.
4   Viewing from the front, apply anti-clockwise pressure on the left-hand camshaft pulley so that the belt will be slack on the tensioner side. Release the tensioner pulley bolts. The spring on the tensioner pulley will automatically tension the belt. Without relaxing pressure tighten the tension pulley bolts. Repeat the procedure on the right-hand pulley and then recheck the timing. Refit the timing belt covers. Ensure that the cover seals are correctly fitted and form a good seal between the covers and the crankcase and inner case extensions. The longer of the cover retaining bolts must be fitted in the outermost position of the left-hand cover. Tighten the bolts to a torque of 0.8 – 1.2 kgf m (6 – 9 lbf ft).

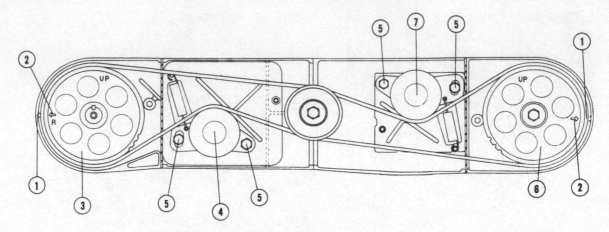

Fig. 1.19 Valve timing – alignment marks

1	Index mark	4	Tensioner pulley	6	Left-hand pulley
2	Arrow mark	5	Bolt	7	Tensioner pulley
3	Right-hand pulley				

53.2a Refit the Woodruff key to each camshaft ...

53.2b ... and slide the driven pulley onto the camshaft end

54 Reassembling the engine/gearbox unit : adjusting the valve clearances and refitting the cylinder head covers

1   Rotate the engine until both valves on No 1 cylinder are fully closed and the 'T-1' mark on the flywheel is aligned with the index marks. With the crankshaft in this position, No 1 piston is at TDC. Check the following valve clearances by placing a feeler gauge of the required size between the valve stem head and the rocker adjuster screw.

No 1 inlet	0.1 mm (0.004 in)
No 1 exhaust	0.13 mm (0.005 in)
No 3 exhaust	0.13 mm (0.005 in)
No 4 inlet	0.1 mm (0.004 in)

If the gap on any valve is incorrect (as will obviously be the case if the adjusters were slackened off prior to cylinder head assembly) loosen the locknut on the adjuster screw and screw the adjuster in or out, as necessary. When adjustment is correct, prevent the screw rotating by using a screwdriver and tighten

the locknut. Re-check the settings. Rotate the engine through 360° until the 'T-1' mark is again aligned with the index marks. No 2 piston is now at TDC on the compression stroke. Check, and adjust where necessary, the valve clearances on the following valves.

No 2 inlet	0.1 mm (0.004 in)
No 2 exhaust	0.13 mm (0.005 in)
No 4 exhaust	0.13 mm (0.005 in)
No 3 inlet	0.1 mm (0.004 in)

2   Check that all the adjuster screw locknuts have been tightened fully. The recommended torque setting is 1.2 – 1.6 kgf m (9 – 12 lbf ft). Refit both cylinder head covers, making sure the rubber seals are correctly positioned, and tighten the retaining bolts to a torque of 0.8 – 1.2 kgf m (6 – 9 lbf ft). Tighten down evenly and in a diagonal sequence. Refit the sealing cap cover to the rear of the left-hand cylinder head/cover, making sure that the gasket has been renewed and that each retaining bolt has a sealing washer fitted beneath its head.

54.1 Check the valve clearances by means of a feeler gauge

54.2 Note the position of the cylinder head rubber seals before fitting the head

55 Reassembling the engine/gearbox unit : refitting the fuel pump

1 Fit the fuel pump to the mounting casting and tighten the two bolts to a torque of 1.8 – 2.2 kgf m (13 – 16 lbf ft).
2 Place the complete assembly in position on the rear of the cylinder head. If difficulty is encountered, rotate the engine so that the pump cam clears the pump arm and the tachometer drive gear meshes easily with the driven gear. Fit and tighten the two retaining bolts to a torque of 1.0 – 1.4 kgf m (7 – 10 lbf ft).

56 Reassembling the engine/gearbox unit: refitting the thermostat assembly

1 Fit new O-rings to the thermostat housing bypass pipes. Lubricate the O-rings with soapy water and fit the pipes to the thermostat housing and manifolds. Insert the hollow locating dowel into the top of the crankcase and fit the O-ring. Fit the complete assembly onto the crankcase, using new gaskets on the two pipe manifolds. Insert the seven retaining screws. Push the pipes firmly into place and at the same time tighten the screws.

2 Insert the thermostat into the main housing, making sure that the vent hole in the front of the unit is positioned uppermost. Refit the thermostat cover, complete with new O-ring, and tighten the retaining bolts to a torque of 2.1 – 2.5 kgf m (15 – 18 lbf ft).
3 Apply a waterproof gasket compound to the threads of the electric fan and water temperature switches and refit them in their respective housings.

57 Reassembling the engine/gearbox unit : refitting the carburettors

1 The carburettors can be refitted only after they have been reassembled into a single unit, the four instruments being attached to the central air box.
2 Place a new O-ring in each of the induction manifolds. Used O-rings should never be employed as they are invariably stretched and may be displaced during bolting up. Place the complete carburettor assembly in position on the top of the engine, locating each manifold flange with the two studs. Fit and tighten the domed retaining bolts with their washers. Reconnect the fuel feed line to the front union on the fuel pump and the feed line to the pulser generator diaphragm casing stub, securing each with their respective retaining clips.

56.1a Fit new O-rings around the thermostat housing dowels ...

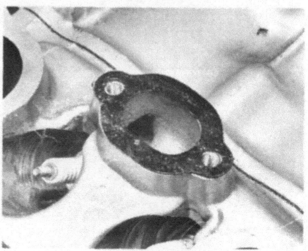

56.1b ... and fit new gaskets on the pipe manifolds ...

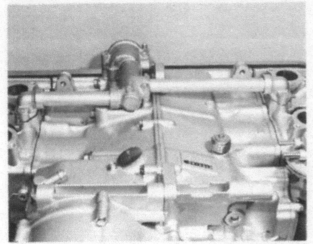

56.1c ... before lowering the complete thermostat assembly into position onto the crankcase

56.2 Position the thermostat vent hole uppermost before fitting the cover

56.3 Refit the electric fan and water temperature switches

58 Reassembling the engine/gearbox unit : refitting the ancillary components

1   Refit the starter motor. Lubricate the O-ring around the motor boss with soapy water to aid insertion into the case. The splined starter motor shaft must engage with the starter drive sprocket as it is inserted. Fit and tighten the two retaining bolts.
2   Screw the oil pressure warning switch into the top of the crankcase, after applying gasket compound to the threads, and tighten it to a torque of 1.0 – 1.4 kgf m (7 – 10 lbf ft).
3   Insert the gauze oil trap into the crankcase with the gauze screen facing the bottom of the engine, making certain that the union locates with the pick-up point in the crankcase. Refit the oil trap cover so that the two pegs on the oil trap locate with the recesses in the cover. Ensure that the cover seal is in good condition before fitting the cover.
4   Place the final drive shaft gaiter in position over the rear of the output shaft and retain it in position on the lip of the engine casting with the retaining spring.
5   To avoid damage occurring to the blades of the water pump impeller during engine installation into the frame, refit the water pump outer casing by first pushing the two locating dowels into their location in the front engine cover. Fit the new self-sealing gasket over the dowels and follow this with the casing. Fit and tighten the four retaining bolts.
6   Finally, reconnect the engine wiring harness to the engine switches, placing the block connectors in position on the crankcase ready for connecting into the main wiring harness.

59 Fitting the engine/gearbox unit into the frame

1   As with engine removal, the use of a trolley jack and the help of two assistants is indispensable for efficient and safe engine refitting. The same system should be adopted as for engine removal, by placing the engine on the jack and wheeling it into position from the left-hand side. Protect the engine sump by inserting a piece of wood between the engine and the jack. During engine refitting a certain advantage is gained in that the engine can be placed on the jack in a position of optimum balance. This was not possible during removal as the exact position was a matter of guesswork.
2   Move the engine towards the frame until it is close enough to reconnect the throttle cables to the throttle pulley. If necessary, move the engine further inwards altering the height of the engine by means of the jack. The throttle pulley will have to be rotated by hand to reconnect the cables easily. Adjust the throttle by following the instructions given in the Routine Maintenance Chapter of this manual.
3   The clutch cable should also be reconnected at this stage as clutch adjustment is made much easier by the access available. Pass the cable through the guide on the rear engine cover and reconnect it to the operating lever. Adjust the clutch in accordance with the instructions given in Routine Maintenance and fit and tighten the clutch cover access cap, having renewed its O-ring.
4   Move the engine inwards until it is in approximately the correct final position, adjusting the engine height as necessary to avoid fouling the frame. Fit the two rear upper mounting plates to the frame by fitting and tightening the four flange bolts, two to each plate. Align the engine mounting lug with the corresponding hole in each mounting plate and fit the securing bolt from the left-hand side of the machine. A plain washer should be fitted beneath the bolt head and the main earth lead between the left-hand mounting plate and engine mounting lug. Fit the plain washer, spring washer and dome nut to the end of the bolt and tighten the nut.
5   Position the fan shroud in the frame and fit and tighten the two shroud to frame attachment bolts followed by the shroud to engine attachment bolts. Note that the shorter of the bolts must be fitted between the shroud and frame.

6 The removable cradle frame member may now be pushed into position over its locating studs and secured in position by fitting the left-hand footrest, with its securing nut and spring washer, the dome nut and spring washer to the rear of the footrest and the dome nut with its plain and spring washers to the uppermost of the two forward studs. It may be found necessary to raise or lower the engine slightly with the jack so that the frame member may be pushed over the studs without undue force being applied. Refit the engine casing to frame member attachment bolt, situated directly above the prop stand pivot, and the corresponding bolt on the opposite side of the machine.

7 If crash bars have been fitted to the machine, they should now be refitted by first pushing the left-hand half of the assembly into position, fitting its mid-point mounting bracket over the lower of the two forward removable frame member mounting studs and loosely fitting the dome nut and spring washer to the stud. The right-hand half of the assembly may then be attached to the left-hand half and the two upper mounting clamps and the one right-hand mounting clamp loosely assembled around the frame downtubes. Align the lower bar mounting holes with those in the frame and engine mounting lugs and gently drift the lower forward mounting bolt into position. Refit the locknut and washer. Note the positioning of the spacer bwtween the left-hand frame lug and crash bar mounting. All the crash bar mounting points may now be tightened.

8 It should be noted that if crash bars have not been fitted to the machine, the one remaining removable frame member securing nut and spring washer, and the lower forward mounting bolt with its locknut and washer, must now be refitted.

9 With all the engine mounting bolts and nuts in position tighten them to the torque values specified below.

12 mm	5.5 – 6.5 kgf m	(40 – 47 lbf ft)
10 mm	3.0 – 4.0 kgf m	(22 – 29 lbf ft)
8 mm	1.8 – 2.5 kgf m	(13 – 18 lbf ft)

10 Lubricate and refit the final drive shaft universal joint over the output shaft and lock it in position with the retaining circlip, pulling the gaiter back over the engine casting to allow access. Check that the circlip is correctly located in the splined shaft and fit the gaiter over the end of the swinging arm. Refit the rear brake stop lamp operating switch spring to both the switch and brake lever.

11 Refit the fuel line to the input stub of the fuel pump and tighten the retaining bracket screw. Loosen and remove the inner bolt from the fuel pump casing flange. Insert the tachometer drive cable, ensuring that the cable end locates correctly, and refit the bolt.

12 Secure the electrical leads to the rear of the fan shroud by means of the built-in clips and reconnect the main wiring leads at the block connectors. Connect the starter motor lead to the terminal and the HT leads to the sparking plugs. Reconnect the choke cable to the operating lever at the carburettor end and secure the cable outer by fitting its securing clamp and tightening the crosshead retaining screw. Check around the engine to see that all electrical wires and control cables have been correctly routed and will not become chafed or damaged during machine operation.

13 Check the condition of the three seals which comprise part of the air filter assembly. If they are perished or have become dry and cracked or split, then they must be renewed. The same applies to the breather pipes of the assembly. Fit the first of the seals to the carburettor central air box followed by the filter housing. Insert the element retainer into the housing and secure it in position with the two retaining bolts. Place the seal around the element retainer and push the element into position. Locate the seal around the top edge of the filter housing and fit the cover, noting that the ducts should face towards the rear of the machine. Secure the cover in position with the wingnut. Finally reconnect the breather pipes from the filter housing to the engine crankcase and from the filter housing to a point clear of

the engine (ie routed forward of the main stand pivot, passing inboard of the final drive shaft housing on the engine rear cover).

14 Turn the forks onto full left-hand lock and lift the complete radiator assembly into position so that it locates over the two upper mounting studs. Fit the electrical harness guard over the left-hand stud and loosely refit the dome nuts and washers onto the studs. With the radiator supported in this position, reconnect the radiator fan electrical leads to the main harness at the block connector and secure the leads by means of the clip provided. Reconnect the radiator top hose to the thermostat housing manifold and tighten the hose clip. Reconnect the pipe from the radiator filler point to the expansion tank, securing it with the wire retaining clip. Refit the two lower mounting bolts and fully tighten both these bolts and the two upper mounting nuts. Refit the water pump manifold to the radiator lower hose and tighten the wire retaining clip. Renew the O-ring in the manifold and secure it to the water pump outer casing by fitting and tightening the two retaining bolts.

15 Refit the gearchange lever onto the splined shaft, making sure the punch marks of the end of the shaft and on the face of the lever are aligned.

16 As with removal, refitting of the exhaust system requires the aid of an assistant. Fit a new gasket into each exhaust port, distorting them very slightly if necessary so that they become oval and remain in the ports. With a person either side of the machine, push the two halves of the system together at the balancer pipe connection. The complete assembly may then be lifted up to locate on the cylinder head studs and align with the rear mounting points. Refit the bolt and nut to each rear mounting and the two flange nuts to each of the four cylinder head manifold connections. Tighten the manifold nuts, followed by the rear mounting nuts and the two bolts retaining the balancer pipe clamp. Take care to tighten the manifold nuts evenly to avoid distortion of the pipe flanges. Note the torque figures given for the exhaust mounting points in the Specifications Section of Chapter 3.

17 Reconnect the leads to the battery terminals and secure the battery in position in its tray with the retaining bracket and bolt. Ensure the vent pipe is routed correctly so that it vents clear of the engine and frame components.

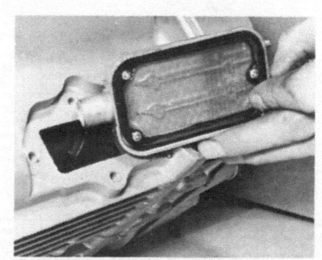

58.3a Insert the gauze oil trap into the crankcase ...

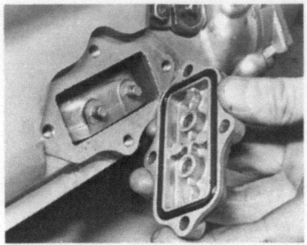

58.3b ... and locate the pegs on the trap with the recesses in the cover

58.4 Retain the final drive shaft gaiter in position with the spring

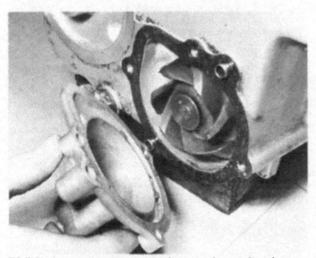

58.5 Fit the water pump outer casing over the two locating dowels

59.3 Reconnect the clutch cable to the operating lever

59.4 Fit the two rear upper mounting plates

60 Starting and running the rebuilt engine

1   Fit a new oil filter into the filter housing and renew the O-ring which forms the housing to engine front cover seal. Refit the housing to the front of the engine, noting that the housing should be fitted so that the aligning tabs align either side of the boss on the water pump cover. Tighten the housing bolt fully to a torque setting of 2.7 – 3.3 kgf m (20 – 24 lbf ft). Refit and tighten the engine oil drain plug, to a torque of 3.5 – 4.0 kgf m (25 – 29 lbf ft), and the coolant drain plug. Ensure each plug has a new sealing washer fitted.
2   Refill the crankcase through the filler orifice to the rear of the fuel pump with the correct specification of engine oil. The engine will accept about 4.0 litres (4.2/3.5 US/Imp quarts) after complete dismantling and reassembly. Check the level through the sight window in the crankcase.
3   Fill the radiator with the recommended coolant. The capacity after complete dismantling is about 3.4 litres (3.6/3.0 US/Imp quarts). Replenish through the radiator filler orifice until the level is at the lower end of the filler neck. The cooling system will need to be bled so that all air is removed. This is accomplished by starting the engine, which should be allowed to run at about 900 rpm for 10 minutes, revving up the engine

for the last 30 seconds to accelerate bleeding. The coolant level will fall, indicating the expulsion of air. Top up the radiator and refit the cap. If necessary, replenish the expansion tank so that the level is just below the upper level mark.

4 Start the engine, and keep it running at a low speed for a few minutes to allow oil pressure to built up and the oil to circulate. If the oil pressure warning lamp is not extinguished, stop the engine immediately and investigate the lack of pressure. The engine may tend to smoke through the exhausts initially, due to the amount of oil used when assembling the components. The excess of oil should gradually burn away as the engine settles down.

5 To check the accuracy of the ignition timing, refer to Chapter 4 and carry out the procedure given in paragraph 2 of Section 8.

6 Check the exterior of the machine for oil leaks or blowing gaskets. Make sure that each gear engages correctly, and that all the controls function effectively, particularly the brakes. This is an essential last check before taking the machine on the road. Refit the dummy fuel tank, seat and sidepanels, using a reversal of the procedure given for removal.

61 Taking the rebuilt machine on the road

1 Any rebuilt machine will need time to settle down, even if parts have been refitted in their original order. For this reason it is highly advisable to treat the machine gently for the first few miles to ensure oil has circulated throughout the lubrication system and that any new parts fitted have begun to bed down.

2 Even greater care is necessary if the engine has been rebored or if a new crankshaft has been fitted. In the case of a rebore, the engine will have to be run-in again, as if the machine were new. This means greater use of the gearbox and a restraining hand on the throttle until at least 500 miles have been covered. There is no point in keeping to any set speed limit; the main requirement is to keep a light loading on the engine, and to gradually work up performance until the 500 mile mark is reached. These recommendations can be lessened to an extent when only a new crankshaft is fitted. Experience is the best guide since it is easy to tell when an engine is running freely.

3 If at any time a lubrication failure is suspected, stop the engine immediately and investigate the cause. If an engine is run without oil, even for a short period, irreparable engine damage is inevitable.

4 When the engine has cooled down completely after the initial run, re-check the various settings, especially the valve clearances. During the run most of the engine components will have settled into their normal working locations.

62 Fault diagnosis : engine

Symptom	Cause	Remedy
Engine will not start	Defective spark plugs	Remove the plugs and lay on cylinder heads Check whether spark occurs when ignition is switched on and engine rotated
	Defective CDI ignition system	Carry out checks listed in Chapter 4
	Valve stuck in guide	Free and clean both stem and valve guide Renew, if binding
	Faulty valve timing	Check and re-set
Engine runs unevenly	Ignition and/or fuel system fault	Check each system independently, as though engine will not start
	Blowing cylinder head gasket	Renew gasket
	Incorrect valve clearance adjustment	Check and adjust
Lack of power	Fault in fuel system or ignition system	See above
	Valve sticking	See above
	Valve seats pitted	Grind in valves
	Faulty valve springs	Renew as a set
	Faulty piston rings	Renew as a set
Engine overheats	Heavy carbon deposit	Decoke engine
	Lean fuel mixture	Adjust carburettors
	Fault in cooling system	See Chapter 2
Heavy oil consumption	Cylinder block in need of rebore	Check for bore wear, rebore and fit oversize pistons if required
	Damaged oil seals	Check engine for oil leaks
	Excessive oil pressure	Check pressure relief valve action

63 Fault diagnosis : clutch

Symptom	Cause	Remedy
Engine speed increases as shown by tachometer but machine does not respond	Clutch slip	Check clutch adjustment for free play at handlebar lever. Check thickness of inserted plates
Difficulty in engaging gears. Gear changes jerky and machine creeps forward when clutch is withdrawn. Difficulty in selecting neutral	Clutch drag	Check clutch adjustment for too much free play. Check clutch drums for indentations in slots and clutch plates for burrs on tongues. Dress with file if damage not too great
Clutch operation stiff	Damaged, trapped or frayed control cable Bent operating pushrod	Check cable and replace if necessary. Make sure cable is lubricated and has no sharp bends Check the pushrod for trueness

64 Fault diagnosis : gearbox

Symptom	Cause	Remedy
Difficulty in engaging gears	Selector forks bent Gear clusters not assembled correctly	Replace Check gear cluster arrangement and position of thrust washers
Machine jumps out of gear	Worn dogs on ends of gear pinions Stopper arms not seating correctly	Replace worn pinions Remove right-hand crankcase cover and check stopper arm action
Gear change lever does not return to original position	Broken return spring	Replace spring

Chapter 2 Cooling system

Contents

Specifications

Cooling system capacity .... 3.4 litres (3.6/3.0 US/Imp quarts)

Coolant

    Specification .... 50/50 solution of distilled water and high quality ethylene glycol antifreeze
Warning: Do not use alcohol based antifreeze

    Boiling point (50/50 mixture):
    Unpressurised .... 107.7°C (226°F)
    Pressurised .... 125.6°C (258°F)

Pressure cap

    Release pressure .... 10.7 – 14.9 psi (0.75 – 1.05 kg/sq cm)

Thermostat

    Type .... Wax pellet
    Begins to open .... 80 – 84°C (176 – 183°F)
    Fully open .... 93 – 96°C (199 – 205°F)
    Valve lift .... Minimum of 8 mm (0.315 in) at 95°C (203°F)

Fan switch

    Fan on .... 98 – 102°C (208 – 216°F)
    Fan off .... 93 – 97°C (199 – 207°F)

Torque wrench settings

	kgf m	lbf ft
Water pump body securing bolts (6 mm)	0.8 – 1.2	6 – 9
Thermostat cover securing bolts (8 mm)	2.1 – 2.5	15 – 18
Thermostat switch	2.4 – 3.2	17 – 23

1 General description

The Honda GL1100 is provided with a car type cooling system which utilises a water/antifreeze coolant to carry away excess energy produced in the form of heat. The cylinders are surrounded by a water jacket from which the heated coolant is circulated by thermo-syphonic action in conjunction with a water pump driven off the front of the oil pump drive shaft. The hot coolant passes upwards through a thermostat housing to the top of the radiator which is mounted on the frame downtubes to take advantage of maximum air flow. The coolant then passes downwards, through the radiator core, where it is cooled by the passing air, and then to the water pump and engine where the cycle is repeated. An electric fan is mounted behind the radiator to aid cooling, when operating conditions demand. The electric fan motor is activated automatically by means of a sensor switch fitted to the thermostat housing. A wax pellet type thermostat is fitted in the system to prevent the flow of coolant through the radiator when the engine is cold, thereby accelerating the speed at which the engine reaches normal working temperature.

The complete system is sealed and pressurised; the pressure being controlled by a valve contained in the spring loaded radiator cap. By pressurising the coolant to between 10.7 and 14.9 psi, the boiling point is raised, preventing premature boiling in adverse conditions. The overflow pipe from the radiator is connected to an expansion tank into which excess coolant is discharged by pressure. The expelled coolant automatically returns to the radiator, to provide the correct level when the engine cools again.

2 Cooling system: draining

1   Place the machine on the centre stand so that it rests on level ground. If the engine is cold, remove the radiator cap in the normal manner by pressing the cap downwards and rotating it in an anti-clockwise direction. If the engine is hot having just been run, place a thick rag over the cap and turn it slightly until all the pressure has been allowed to disperse. A rag must be used to prevent escaping steam from causing scalds to the hand. If the cap were to be removed suddenly (or allowed to be forced off), the drop in pressure could allow the water to boil violently and be expelled under pressure from the filler neck. Apart from burning the skin the water/antifreeze mixture will damage paintwork.

2   Place a receptacle below the front of the engine into which the coolant can be drained. The container must be of a capacity greater than the volume of coolant which is 3.4 litres (3.6/3.0 US/Imp quarts). If the coolant is to be re-used, ensure that the container is perfectly clean. Remove the drain plug from the water pump casing and allow the coolant to drain completely.

3   To complete the draining of the system it will be necessary to detach the expansion tank from the machine and invert it to allow any coolant in the tank to drain from the filler point. To achieve this, remove the dummy fuel tank as described in Chapter 3, detach the tubes from the tank stubs by releasing the wire clips, unscrew and remove the single retaining bolt from the lower bracket of the tank, unscrew the filler cap and pull the tank away from the machine at the same time releasing it from the front mounting stay. Once drained, the tank should be refitted using a reverse of the removal procedure.

4   Note that the recommended interval at which the coolant should be renewed is 24 000 miles or every two years.

3 Cooling system: flushing

1   After extended service the cooling system will slowly lose efficiency, due to the build up of scale, deposits from the water and other foreign matter which will adhere to the internal surfaces of the radiator and water channels. This will be particularly so if distilled water has not been used at all times. Removal of the deposits can be carried out easily, using a suitable flushing agent in the following manner.

2   After allowing the cooling system to drain, refit the drain plug and refill the system with clean water and a quantity of flushing agent. Any proprietary flushing agent in either liquid or dry form may be used, providing that it is recommended for use with aluminium engines. NEVER use a compound suitable for iron engines as it will react violently with the aluminium alloy. The manufacturer of the flushing agent will give instructions as to the quantity to be used.

3   Run the engine for ten minutes at operating temperatures and drain the system. Repeat the procedure TWICE and then again using only clean cold water. Finally, refill the system as described in the following Section.

4 Cooling system: filling

1   Before filling the system, check that the sealing washer on the drain plug is in good condition and renewed if necessary. Fit and tighten the drain plug and check and tighten all the hose clips.

2   Fill the system slowly to reduce the amount of air which will be trapped in the water jacket. When the coolant level is up to the lower edge of the radiator filler neck, run the engine for about 10 minutes at 900 rpm. Increase engine revolutions for the last 30 seconds to accelerate the rate at which any trapped air is expelled. Stop the engine and replenish the coolant level again to the bottom of the filler neck. Refill the expansion tank up to the 'Full' level mark. Refit the radiator cap, ensuring that

it is turned clockwise as far as possible.

3   Ideally, distilled water should be used as a basis for the coolant. If this is not readily available, rain water, caught in a non-metallic receptacle, is an adequate substitute as it is de-ionised and contains only limited amounts of mineral impurities. If absolutely necessary, tap water can be used, especially if it is known to be of the soft type. Using non-distilled water will inevitably lead to early 'furring-up' of the system and the need for more frequent flushing. The correct water/antifreeze mixture is 50/50; do not allow the antifreeze level to fall below 40% as the anti-corrosion properties of the coolant will be reduced to an unacceptable level. Anti-freeze of the enthylene glycol based type should always be used. Never use alcohol based anti-freeze in the engine.

4.2 Refill the expansion tank to the FULL level mark

5 Radiator: removal, cleaning, examination and refitting

1   Drain the radiator as described in Section 2 of this Chapter.

2   Disconnect the top hose at the thermostat manifold by loosening the screw clip. Because the lower hose is of a short length only, it is easier to disconnect it by detaching the manifold from the water pump casing. The manifold is retained by two bolts. The screw clip can now be loosened and the manifold pulled from the lower hose. Disconnect the pipe connecting the expansion tank to the radiator filler point by removing the wire retaining clip and pulling the pipe end off the radiator stub.

3   Remove the radiator lower mounting bolts. Support the radiator with one hand and remove the top mounting nuts. Ease the radiator forwards and disconnect the electrical leads to the electric fan by pulling the 'block' connector apart. Turn the front forks onto full left-hand lock and carefully lift the radiator from position. Take care not to damage the radiator fins.

4   Remove the fan shroud and fan as a unit by removing the three mounting bolts. Detach the radiator guard from the front of the radiator, where it is retained by four domed nuts, and remove the shrouds from each side of the radiator by unscrewing the securing nuts.

5   Remove any obstructions from the exterior of the radiator core, using an air line. The conglomeration of moths, flies, and autumnal detritus usually collected in the radiator matrix severely reduces the cooling efficiency of the radiator.

6   The interior of the radiator can most easily be cleaned while the radiator is in-situ on the motorcycle, using the flushing procedure described in Section 3 of this Chapter. Additional flushing can be carried out by placing the hose in the filler neck and allowing the water to flow through for about ten minutes. Under no circumstances should the hose be connected to the filler neck mechanically as any sudden blockage in the radiator

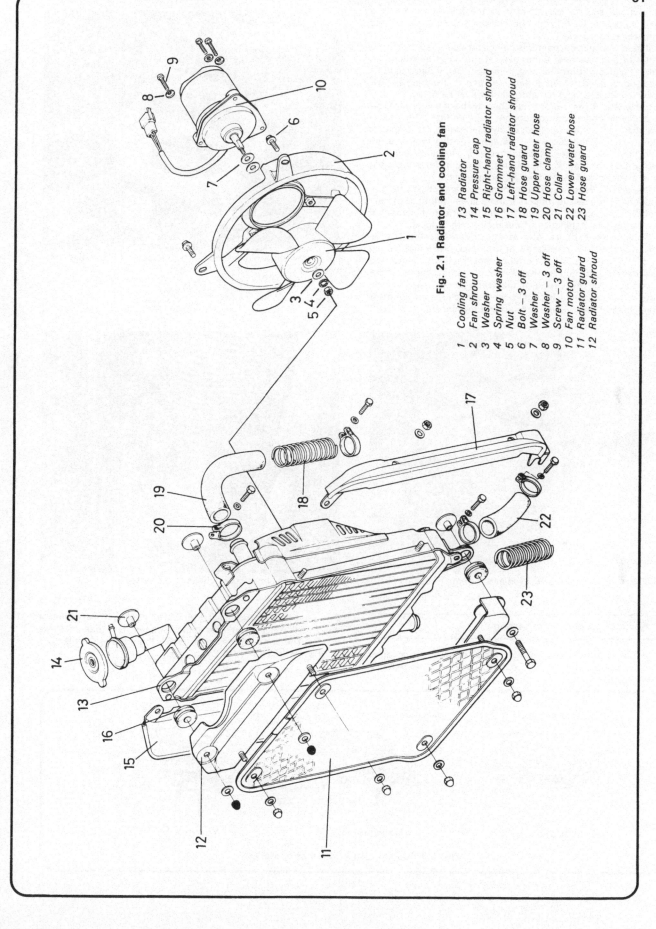

Fig. 2.1 Radiator and cooling fan

1 Cooling fan
2 Fan shroud
3 Washer
4 Spring washer
5 Nut
6 Bolt – 3 off
7 Washer
8 Washer – 3 off
9 Screw – 3 off
10 Fan motor
11 Radiator guard
12 Radiator shroud
13 Radiator
14 Pressure cap
15 Right-hand radiator shroud
16 Grommet
17 Left-hand radiator shroud
18 Hose guard
19 Upper water hose
20 Hose clamp
21 Collar
22 Lower water hose
23 Hose guard

outlet would subject the radiator to the full pressure of the mains supply (about 50 psi). The radiator should not be tested to greater than 15 psi.

7   If care is exercised, bent fins can be straightened by placing the flat of a screwdriver either side of the fin in question and carefully bending it into its original shape. Badly damaged fins cannot be repaired. If bent or damaged fins obstruct the air flow more than 20%, a new radiator will have to be fitted.

8   Generally, if the radiator is found to be leaking, repair is impracticable and a new component must be fitted. Very small leaks may sometimes be stopped by the addition of a special sealing agent in the coolant. If an agent of this type is used, follow the manufacturer's instructions very carefully. Soldering, using soft solder may be efficacious for caulking large leaks but this is a specialised repair best left to experts.

9   Inspect the four radiator mounting rubbers for perishing or compaction. Renew the rubbers if there is any doubt as to their condition. The radiator may suffer from the effect of vibration if the isolating characteristics of the rubber are reduced.

10   Once serviced, the radiator assembly may be refitted to the machine by using a reversal of the removal procedure, noting the following points. Do not omit to refit the electrical harness guard beneath the top left-hand radiator retaining nut. Check the condition of the O-ring in the water pump manifold and renew it if necessary. Check all hose connections for tightness before refilling the system and running the engine.

5.2 The overflow pipe is secured to the radiator filler point stub by a wire clip

5.9 Check the condition of the radiator mounting rubbers

5.10 Check all hose connections for tightness

6   Radiator pressure cap: testing

1   If the valve or valve spring in the radiator cap becomes defective the pressure in the cooling system will be reduced, causing boiling over.

2   A special pressure cap tester will be needed to determine at which point the valve lifts; most garages have one of these testers. The correct pressure at which the valve will lift is 10.7 – 14.9 psi (0.75 – 1.05 kg/sq cm).

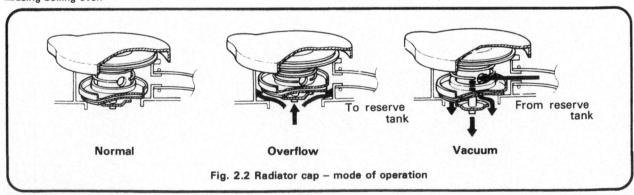

Normal   Overflow   Vacuum

To reserve tank   From reserve tank

Fig. 2.2 Radiator cap – mode of operation

7 Thermostat: removal, testing and refitting

1   The thermostat is so designed that it remains in the closed position when it is in a normal cold condition. If the thermostat malfunctions, it will remain closed even when the engine reaches normal working temperature. The flow of coolant will be impeded so that it does not pass through the radiator for cooling and consequently the temperature will rise abnormally, causing boiling over.

2   If the performance of the thermostat is suspect, remove it from the machine as follows and test it for correct operation. Drain the coolant and remove the radiator as previously described. Remove the thermostat manifold, which is retained by two bolts and lift the thermostat from position.

3   Examine the thermostat visually before carrying out tests. If it remains in the open position at room temperature, it should be discarded.

4   To test the thermostat, suspend it by a piece of wire in a pan of cold water. Place a thermometer in the water so that the bulb is close to the thermostat. The type of thermometer used when preparing fruit preserves is ideal. Heat the water, noting when the thermostat opens and the temperature at which the thermostat is fully open. If the performance is at variance with the following table the thermostat should be renewed.

Valve begins to open   80 – 84°C (176 – 183°F)
Valve fully open   93 – 96°C (199 – 205°F)

Heat the thermostat for about 5 minutes at 97°C (207°F) and measure the valve lift which should be 8 mm (0.315 in).

5   Refit the thermostat in its housing so that the vent hole is in the uppermost position. Refit the manifold, checking that the O-ring is in good condition and correctly located in the manifold recess. Fit and tighten the two securing bolts. Refit the radiator and refill the system with coolant.

8 Water pump: removal, renovation and fitting

1   Malfunction of the water pump requires renewal of the component as repair is impracticable. The most likely faults are looseness and damage to the impeller or leakage of the oil/water seal leading to failure of the pump bearings.

2   Leakage of the seal will allow water to find its way into the lubrication system or oil to seep into the coolant. In either event the pump should be renewed immediately, removal and refitting being carried out as follows.

3   Drain the cooling system and also the engine oil. Remove the water pump casing, which is retained by four bolts. The lower hose manifold need not be detached. The pump casing is located on two dowels and owing to this and the type of gasket used, may be difficult to remove. A soft-nosed mallet may be used to tap gently the case from position. Examination of the impeller can be made at this stage without further dismantling.

4   Remove the radiator and unscrew the centre bolt and remove the oil filter housing. Loosen the engine front cover screws evenly and detach the cover. Note the position of the various O-rings, hollow dowels and collars.

5   The water pump is retained from inside the cover by three bolts. Remove the bolts and drift the pump from position. Do not tap the shaft end or damage will result. Check the bearings for wear and check also for any evidence of leakage past the shaft. Seepage of oil or water thought to be caused by a faulty pump seal may be caused by faulty or omitted O-rings though this is not very likely. As mentioned previously, repair of this component is impracticable and in any event, spares are not available over the counter.

6   Fit the water pump by reversing the order given for removal, noting the following points. All O-rings must be renewed if they are found to be damaged or in poor condition. Do not omit to fit any O-rings. Tighten the three water pump securing bolts to a torque of 0.8 – 1.2 kgf m (6 – 9 lbf ft). Tighten the pump cover and engine front cover securing bolts evenly and in a diagonal sequence. Fit new gaskets to both the pump cover and engine front cover mating surfaces.

9 Fan motor: testing

Refer to Chapter 7, Section 25.

10 Sensor switches: testing

Refer to Chapter 7, Sections 24 and 25.

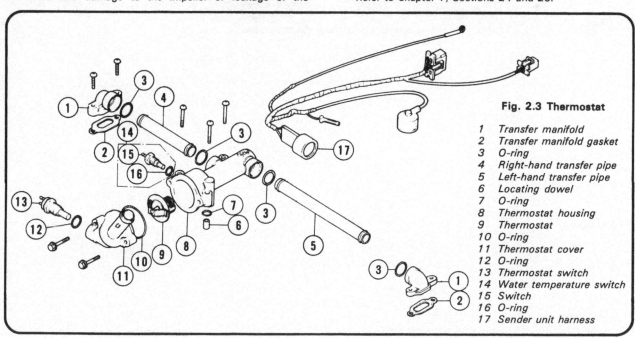

Fig. 2.3 Thermostat

1   Transfer manifold
2   Transfer manifold gasket
3   O-ring
4   Right-hand transfer pipe
5   Left-hand transfer pipe
6   Locating dowel
7   O-ring
8   Thermostat housing
9   Thermostat
10 O-ring
11 Thermostat cover
12 O-ring
13 Thermostat switch
14 Water temperature switch
15 Switch
16 O-ring
17 Sender unit harness

7.5 Note the vent hole in the thermostat case

8.6a Renew both water pump body O-rings ...

8.6b ... fit the pump and secure it with the three retaining bolts

11 Fault diagnosis: cooling system

Symptom	Cause	Remedy
Overheating	Insufficient coolant in system	Top-up reservoir
	Radiator core internally blocked	Flush out system
	Radiator core externally blocked	Remove and clean radiator
	Hoses collapsed blocking flow	Remove and fit new hose(s)
	Thermostat not opening correctly	Remove and fit new thermostat
	Blowing cylinder head gasket (water/steam being expelled into reservoir and out of breather under pressure)	Remove cylinder head and fit new gasket
	Low percentage of antifreeze	Drain and refill with 50/50 mixture
	See also mechanical causes fault diagnosis, Chapter 1.	
Underheating	Thermostat jammed open	Remove and renew
	Incorrect grade of thermostat	Fit thermostat in correct range
	Thermostat omitted from system	Check and fit correct thermostat
Loss of coolant	Loose hose clips	Tighten clips
	Hose leaking	Renew hose
	Radiator core leaking	Small leaks, use anti-leak compound
		Large leaks, remove for repair or renewal
	Thermostat, water pump casing or manifold gasket leaking	Renew gasket
	Radiator cap valve spring worn	Test cap and renew
	Blown cylinder head gasket (water/steam being forced out of system under pressure	Remove cylinder head and renew gasket
	Cylinder wall or head cracked	Dismantle engine, renew damaged parts or have expert repair made

Chapter 3 Fuel system and lubrication

Contents

Specifications

Fuel tank

Capacity	20 litres (5.3/4.4 US/Imp galls)
Reserve capacity	4 litres (1.1/0.9 US/Imp galls)

Carburettors

Make	Keihin
Type	VB
Venturi bore size	30 mm (1.80 in)
ID no	VB48A
Main jet	145
Jet needle	64A
Float level	15.5 mm (0.61 in)
Pilot screw initial setting	$1\frac{1}{4}$ turns out
Idle speed	950 ± 100 rpm
Fast idle speed	2500 − 3500 rpm

Fuel pump

Type	Cam operated, diaphragm

Engine oil capacity

After draining	3.2 litres (3.4/2.8 US/Imp quarts)
After dismantling	4.0 litres (4.2/3.5 US/Imp quarts)

Oil pumps

Main and clutch scavenge pump type	Earls trochoid
Rotor to body side clearance:	
Main pump	0.02 − 0.07 mm (0.0008 − 0.0028 in)
Service limit	0.12 mm (0.0047 in)
Clutch pump	0.02 − 0.10 mm (0.0008 − 0.0039 in)
Service limit	0.12 mm (0.0047 in)
Rotor to body radial clearance:	
Both pumps	0.15 − 0.21 mm (0.0059 − 0.0083 in)
Service limit	0.41 mm (0.0161 in)
Inner to outer rotor radial clearance:	
Both pumps	0.15 mm (0.0059 in)
Service limit	0.35 mm (0.138 in)

Torque wrench settings

	kgf m	lbf ft
Fuel pump securing bolts (8 mm)	1.8 – 2.2	13 – 16
Fuel pump cover bolts (6 mm)	1.0 – 1.4	7 – 10
Oil filter centre bolt	2.7 – 3.3	20 – 24
Oil drain bolt	3.5 – 4.0	25 – 29
Oil pressure switch	1.0 – 1.4	7 – 10
Exhaust pipe nuts (8 mm)	1.5 – 2.0	11 – 14
Exhaust silencer mounting bolt (10 mm)	3.5 – 4.5	25 – 33
Exhaust silencer joint band (8 mm)	2.0 – 2.4	11 – 17

1 General description

The fuel system fitted to the Honda GL 1100 is similar to that used on most modern motorcycles. It is, however, unusual in the three following particulars. The fuel is contained within a tank fitted below the dualseat and forward of the rear mudguard, the normal position of the fuel tank being occupied by a dummy tank containing the air filter and electrical components.

An indication of the quantity of fuel being carried is given by a fuel gauge, contained in the handlebar mounted instrument console.

Fuel is fed from the tank to the carburettors by means of a diaphragm type fuel pump, fitted to the rear of the right-hand cylinder head and operated by a cam lobe extending from the valve operating camshaft. The remainder of the fuel system follows normal motorcycle practice.

Four 30 mm Keihin constant vacuum carburettors are mounted over the top of the engine, on a common cast aluminium air box and feeding individual cylinders through separate inlet manifolds. The air box is attached to an air filter box within the dummy fuel tank, containing a corrugated paper type element. The carburettors are interconnected by a control rod and are operated from a throttle twist grip by push-pull cables around a single pulley.

The rearmost of the two right-hand carburettors has an accelerator pump mounted on its underside. Although a fuel pump is fitted, a fuel tap is incorporated in the system, attached to the fuel tank, from where it supplies the fuel through a hose and filter to the intake side of the pump. The tap incorporates a 'Reserve' position to enable an additional supply of 4 litres (1.1/0.9 US/Imp galls) of fuel to be available when the main supply runs out. For cold starting, a cable-operated choke is fitted, attached to the rear right-hand carburettor and inter-connected with the remaining three instruments. The choke operating knob is located to the left of the instrument console, just below the speedometer. Lubrication is by the wet sump principle in which oil is delivered, under pressure, from the sump by a mechanical pump to the working parts of the engine. Oil is returned to the sump by gravity and by a secondary pump which scavenges oil trapped in the clutch housing. The two pumps are driven by a shared shaft running the length of the engine and driven by a duplex chain taken off a sprocket to the rear of the clutch outer drum. Both pumps are of the trochoid rotating vane type. Oil is picked up by the main pump, through a gauze strainer in the sump and passed under pressure through a full flow oil filter fitted with a paper element. The engine oil supply is also shared by the gearbox and primary drive.

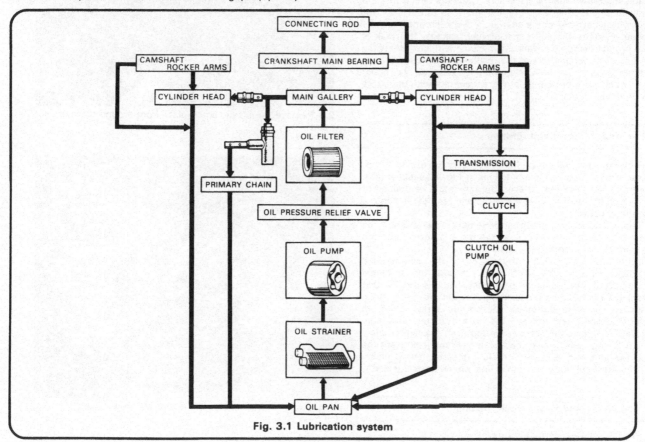

Fig. 3.1 Lubrication system

2 Fuel tank: removal and refitting

1   The fuel tank is retained in position by a single bolt through a bracket at the upper rear edge of the tank, and is supported on two rubber cushions by brackets projecting from the frame.

2   Removal is unnecessary except when the tank itself has become damaged or attention to the rear portion of the frame is required. Owing to the unusual shape of the tank and its close proximity to the frame members and ancillary components bolted to the frame, removal is a laborious process.

3   Commence tank removal by detaching the dummy fuel tank, the dualseat and the side panels. Follow this by removing the rear wheel, mudguard and any other component which may obstruct the tank as it is lifted out towards the rear. Disconnect the fuel line at the fuel pump and allow the tank to drain. Disconnect the two leads from the fuel gauge sensor unit and remove the plastic catch plate from around the fuel filler point.

4   Close inspection of the machine will reveal why removal of the tank is so difficult, and should not be undertaken unless absolutely necessary.

5   Take great care when draining fuel to place the machine in a well ventilated space where there is no chance of a build-up of fuel vapour occurring with the subsequent risk of a fire or explosion. Always drain the fuel into a clean metal container of the correct capacity; never use a plastic container. On completion of draining the fuel, fit the cap to the container and store it in a well ventilated open space. Never smoke or expose the machine to any kind of naked flame when carrying out this operation. Once the tank is removed, ensure the filler cap is tightly fitted and the fuel tap turned to the 'Off' position before storing the tank in conditions similar to those previously described for the fuel container.

6   Refitting the tank is essentially a reversal of the removal procedure. Inspect the fuel feed pipes for deterioration or damage and renew them if necessary. Once the tank is fitted in the machine, start the engine and allow it to run at tick-over speed whilst carrying out a thorough check around the pipe connection points for any signs of fuel leakage. If a leak is found, it must be cured before the machine is ridden otherwise there is a considerable risk of fire resulting in serious injury to the rider. Ensure the tank is properly located on its rubber cushions and does not come into direct contact with the frame or cycle components.

3 Fuel tap: removal and refitting

1   Fuel tap removal is seldom necessary unless the joint between the tap and tank starts to leak or the valve rubber fails. Before removing the tap or lever, the tank must be drained of fuel. Detach the feed hose at the pump and feed the fuel into a clean metal container.

3   Renewal of the complete tap assembly is necessary if the valve rubber fails. This is because the tap is a sealed unit and only available as a complete assembly.

3   In the case of a leak occurring between the tap and tank, check tighten the two screws that pass through the tap flange into the tank. If they are found to be loose then it is possible that tightening them will stop the leak. If this is not the case and the screws are properly tightened, loosen the screw clip from around the feed hose and pull the hose from the tap union. Remove the screws and detach the tap from the tank. Renew the two O-rings which seal the tap flange/tank joint. Refit the tap to the tank and reconnect the fuel hose to the tap union. Start the machine and carry out a final check for fuel leaks.

4 Fuel gauge and float switch: testing

Refer to Chapter 7, Section 26.

2.1 The fuel tank is retained by a single bolt

2.3 Remove the plastic catch plate from around the fuel filler point

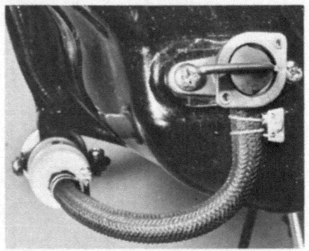
3.3 The fuel tap is secured to the tank by two screws

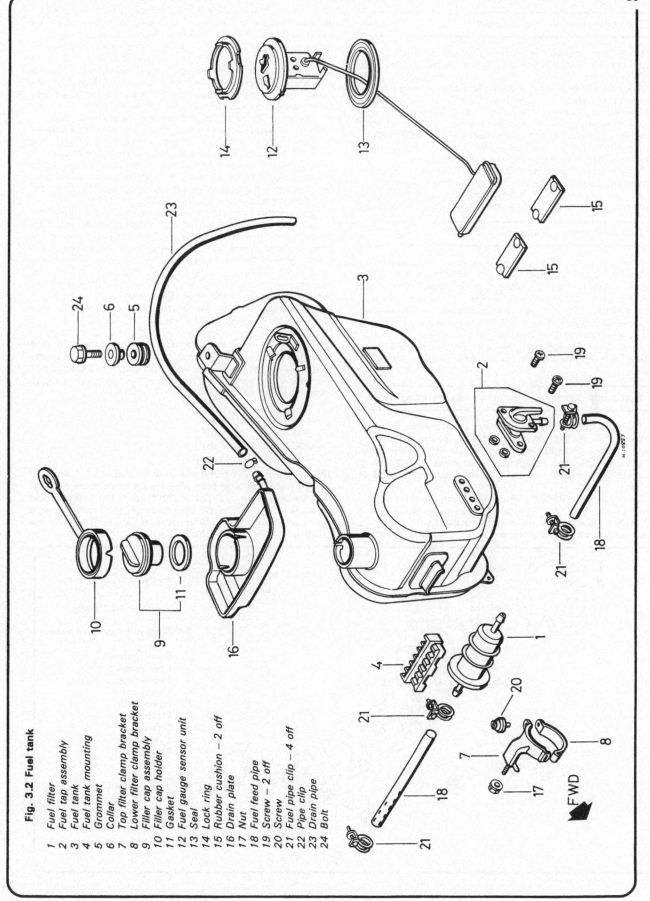

Fig. 3.2 Fuel tank

1 Fuel filter
2 Fuel tap assembly
3 Fuel tank
4 Fuel tank mounting
5 Grommet
6 Collar
7 Top filter clamp bracket
8 Lower filter clamp bracket
9 Filler cap assembly
10 Filler cap holder
11 Gasket
12 Fuel gauge sensor unit
13 Seal
14 Lock ring
15 Rubber cushion – 2 off
16 Drain plate
17 Nut
18 Fuel feed pipe
19 Screw – 2 off
20 Screw
21 Fuel pipe clip – 4 off
22 Pipe clip
23 Drain pipe
24 Bolt

FWD

H: 10657

5 Fuel filter: removal and renewal

1   In addition to two fuel strainers which are fitted in the bottom of the fuel tank, a fuel filter is fitted midway in the tank to fuel pump line.

2   It is recommended that the filter is renewed every 24 000 miles or two years. The filter is a sealed unit and therefore cleaning is impracticable. A new unit must be fitted.

3   The filter is secured by a clamp bracket, which is retained by one bolt to the lower seam of the fuel tank. Remove the filter by separating the clamp and pulling the fuel feed hose off each end of the filter, after loosening the screw clips.

4   Note that the new filter unit must be fitted with the arrow on its casing pointing in the direction of fuel flow, ie towards the fuel pump. On completion of fitting the unit, start the engine and check for fuel leaks at the disturbed fuel hose connections.

6 Fuel feed pipes: examination

1   Inspect the two main fuel feed pipes at intervals for chafing or other damage. Both pipes are made from heavy duty, fabric reinforced rubber and are unlikely to fail in normal service, but may become damaged if the hose clips are overtightened.

2   If persistent fuel starvation is encountered, inspect the inner surface of each pipe. Occasionally, the inner liner of the pipe will come away, causing blockage.

3   It is recommended that the fuel line be renewed every 24 000 miles or two years, to preclude possible failure in service.

7 Fuel pump: removal, examination and refitting

1   The fuel pump is of the lever-operated diaphragm type and is driven off a cam lobe extension of the right-hand camshaft. If foreign matter finds its way into the pump, it may cause malfunctions of the intake and output valves. If pump performance is suspect, it should be removed and cleaned as follows.

2   Unscrew the two bolts which retain the pump mounting casing to the rear of the right-hand cylinder head. Pull the tachometer cable from position in the casing. Remove the two pump flange bolts and detach the pump complete. Note the thick plastic spacer pieces on each casing mounting flange.

3   Unscrew the five cross head screws from the top of the pump and lift off the cap. Remove the valve cage retainer plate which is retained by a single central crosshead screw, and lift out the valves. Clean all the components thoroughly in petrol and inspect them for signs of damage or deterioration. Fit serviceable non-return valves so that when viewed in the pump's normal working position the valve on the input side faces downwards and the output valve faces upwards.

4   Complete failure of the fuel pump is usually caused by the rubber diaphragm perishing and finally splitting. This can be checked without removing the fuel pump body from the machine, by detaching the pump cap. Failure of the diaphragm will be evident on inspection. Owing to modern design trends, the diaphragm cannot be detached from the pump as a whole. Nor are spare diaphragms available as spares. The complete unit must be renewed.

5   Reassemble the pump in the reverse order given for dismantling. Inspect the plastic spacer pieces for deterioration or damage and renew if necessary. Refit the fuel pump to the mounting casing and tighten the two bolts to a torque of 1.8 – 2.2 kgf m (13 – 16 lbf ft). Place the complete assembly in position on the rear of the cylinder head. If difficulty is encountered, rotate the engine so that the pump cam clears the pump arm and the tachometer drive gear meshes easily with the driven gear. Fit and tighten the two retaining bolts to a torque of 1.0 – 1.4 kgf m (7 – 10 lbf ft).

6   If required, the performance of the fuel pump can be checked as follows. Detach the hose from the output side and reconnect the hose via a T-piece. Attach a fluid pressure gauge to the T-piece by means of a length of hose. Run the engine at the following speeds and compare the pressure readings recorded with those below

500 rpm	2.4 psi
900 rpm	2.3 psi
5000 rpm	2.0 psi

The flow rate of the pump can also be checked by using the same T-piece, the hose from which leads into a measuring jug. Run the engine at a speed of 3000 rpm. The pump should supply 0.45 litre (0.95/0.80 US/Imp pint) per minute. The tests described above are usually unnecessary as examination of the pump will reveal any irregularities.

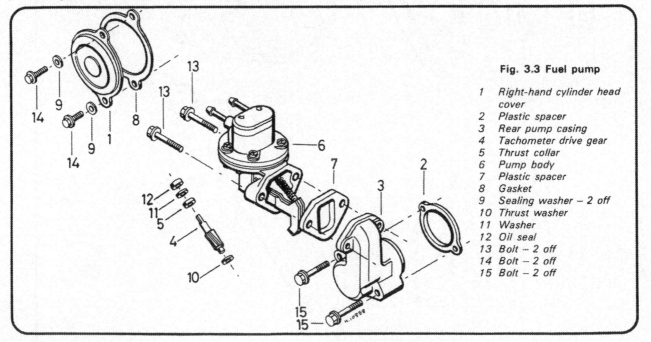

Fig. 3.3 Fuel pump

1   Right-hand cylinder head cover
2   Plastic spacer
3   Rear pump casing
4   Tachometer drive gear
5   Thrust collar
6   Pump body
7   Plastic spacer
8   Gasket
9   Sealing washer – 2 off
10 Thrust washer
11 Washer
12 Oil seal
13 Bolt – 2 off
14 Bolt – 2 off
15 Bolt – 2 off

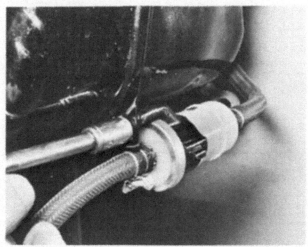

5.3 Remove the filter bracket retaining bolt to free the filter

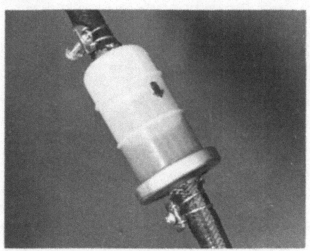

5.4 The arrow on the filter casing must face the direction of fuel flow

7.3 The fuel pump valve cage retainer plate is secured by a single screw

7.4 Failure of the pump diaphragm will be evident on inspection

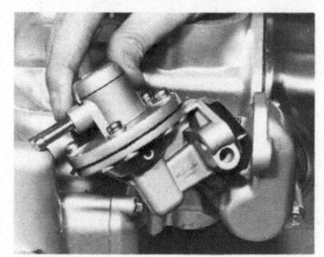

7.5 Place the complete pump assembly in position on the rear of the cylinder head

8 Carburettors: removal from the machine

1 Remove the seat by unscrewing the Allen bolt situated each side at the base of the pillion seat, moving the seat backwards to detach it from the forward frame mounting and, lifting the front of the seat clear of the rear of the dummy fuel tank so that it can be pulled forwards and away from the machine. The dummy fuel tank may now be detached from its frame mounting points by unscrewing the four securing bolts, two at the front of the tank and two at the rear. Open the flaps on the tank top and lift out the tool tray. The dummy tank may now be carefully lifted up, bearing it slightly to the right so that it clears the components located within it.

2 Remove the air filter cover by unscrewing the single retaining wingnut and lifting the cover clear of the filter housing, along with its seal. The filter element and seal may be withdrawn from the housing. Remove the element retainer by unscrewing its two securing bolts and pulling it out of the filter housing. Finally, unclip the breather pipes from the housing stubs and lift the housing clear of the carburettor manifold inlet, noting the sealing ring at the base of the housing.

3 Detach the HT leads from their retaining clips on the carburettor stay plates and pull the suppressor caps off the

sparking plugs. Disconnect the choke cable from its clamp located just to the rear of the rear right-hand carburettor; the outer cable may be freed simply by unscrewing the crosshead retaining screw. Free the cable nipple from the operating lever.

4   Loosen the hose clip from the fuel output pipe at the fuel pump and pull the pipe off the union. Disconnect the control tube to the vacuum diaphragm of the CDI pulser generator at the carburettor body.

5   Remove the eight domed bolts, two each of which retain each carburettor manifold to the cylinder heads. Using a rawhide mallet, tap the complete assembly upwards, until it is free of the cylinder heads. Lift the assembly upwards and move it out to the left-hand side of the machine until the throttle cable attachments are exposed.

6   Release the throttle cables by loosening the cable adjuster nuts at the engine end to allow the adjusters to pass fully down through their guides. The cable nipples may now be released from the operating linkage by opening or closing the throttle

twistgrip, so that either one of the cables is slack, and using a pair of long-nosed pliers to release the nipple of the slackened cable from the operating pulley. To free the cable ends from their adjuster guides, fully undo the adjuster nuts and pull the adjusters up through the guides so that the cable inners may be moved sideways to clear the guide slots. The carburettor assembly may now be pulled clear of the machine.

9   Carburettors: dismantling, examination and reassembly

1   Removal of the carburettors from the air box need not take place unless the instruments are to be renewed or the air box itself requires renewal. Follow the procedure below for carburettor removal.

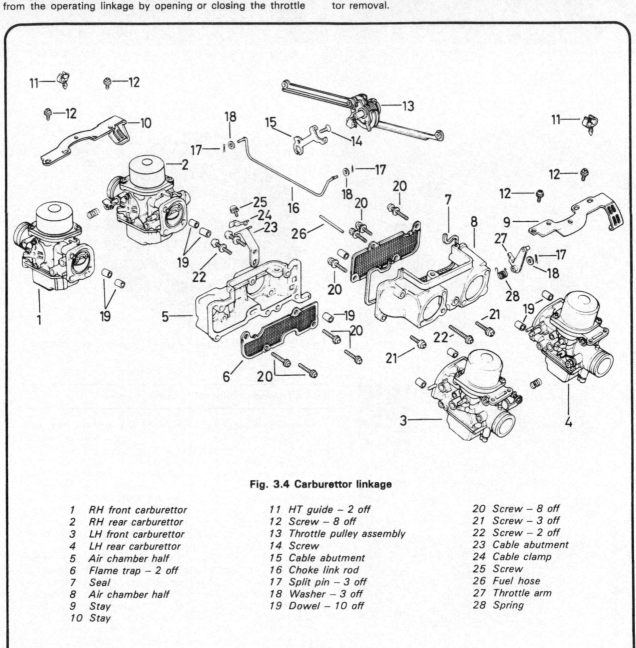

Fig. 3.4 Carburettor linkage

1 RH front carburettor	11 HT guide – 2 off	20 Screw – 8 off
2 RH rear carburettor	12 Screw – 8 off	21 Screw – 3 off
3 LH front carburettor	13 Throttle pulley assembly	22 Screw – 2 off
4 LH rear carburettor	14 Screw	23 Cable abutment
5 Air chamber half	15 Cable abutment	24 Cable clamp
6 Flame trap – 2 off	16 Choke link rod	25 Screw
7 Seal	17 Split pin – 3 off	26 Fuel hose
8 Air chamber half	18 Washer – 3 off	27 Throttle arm
9 Stay	19 Dowel – 10 off	28 Spring
10 Stay		

2   Remove the split-pins and plain washers from the link pivots at the end of the throttle control arms on No 3 and No 4 carburettors. Unscrew and remove the linkage centre bolt, along with the complete linkage assembly, from the carburettors. Alternatively, leave the centre bolt in position and detach the control arms from the throttle cable pulley by removing the remaining split-pins and washers. The former method may well prove to be the more complicated of the two because the main throttle return spring will unwind when the linkage is removed. If the spring is removed be sure to note its fitted position and the number of turns it unwinds on removal.

3   Disconnect the choke control rod link arm from No 3 carburettor by removing the two split pins and plain washers. Remove the five cross-head screws which hold the two halves of the air box together. These screws may be very tight, so great care must be taken to use a screwdriver of the correct type and size otherwise damage will occur to the screwheads. Retain the small lockwashers which are located beneath the screw heads and note the positions of the two longer screws and the choke cable guide for reference when refitting.

4   Pull the two halves of the air box apart. Do not prise the air box halves apart using screwdrivers or other levers as the sealing ring or mating surfaces may be damaged. The two pairs of carburettors can now be removed from their respective air box halves, using an identical procedure for each as follows.

5   Remove the four crosshead screws that secure the gauze screen to the inside of the air box half. These screws also secure the two carburettors to the air box half. Lift the screen out of the air box and separate the air box from the two carburettors. Detach the chromed stay plate from the two carburettors by removing the four crosshead retaining screws. The carburettors can now be separated, taking care to hold them both horizontally to prevent damage occurring to the fuel joint pipes.

6   The manifold with its rubber intake adaptor may now be detached from each carburettor by loosening the clamp retaining screw. Do not try to remove the intake adaptors from their manifolds because the two components are bonded together. Avoid excessive twisting between the two components because this may alter their alignment.

7   It is strongly advised that each carburettor be dismantled and reassembled separately, to prevent accidentally interchanging the components. Dismantle and examine each carburettor, following an identical procedure as described below. Before any dismantling work takes place, drain out any residual fuel and clean the outside of the instruments thoroughly. It is essential that no debris finds its way inside the carburettors.

8   Remove the two vacuum chamber cover retaining screws and lift the cover from position along with the plastic seal ring. Carefully lift the vacuum piston out of the carburettor body and place it on a clean worksurface. Remove the helical spring and the plug from the top of the piston, noting the O-ring on the plug. Insert a small screwdriver into the hole in the top of the piston and unscrew the needle retaining screw. Invert the piston and catch the screw and needle as they fall from the piston. Before placing the piston to one side inspect it for signs of wear or damage and check that it moves freely in the carburettor body. Check also for wear between the piston needle and the jet in which it slides. Wear should not occur between these two components until they have been in service for a considerable amount of time but once it does occur, petrol consumption will begin to increase. Never interchange the piston or cap of one carburettor with that of another.

9   Remove the screw and plate from the carburettor body upper chamber to reveal the air jet located beneath it. This jet and the two similarly positioned opposite it, are not removable and must therefore be cleaned in situ using the process described in paragraph 17 of this Section.

10   Invert the carburettor and remove the float chamber retaining screws. Lift the float chamber from position and note the chamber sealing ring. This need not be disturbed unless it is damaged. The two floats, which are interconnected, can be lifted away after displacing the pivot pin. The float needle is attached to the float tang by a small clip. Detach the clip from the tang and store the needle in a safe place.

11   Check the condition of the floats. If they are damaged in any way, they should be renewed. The float needle and needle seating will wear after lengthy service and should be inspected carefully. Wear usually takes the form of a ridge or groove, which will cause the float needle to seat imperfectly. Remove the needle seat from the carburettor body and inspect the filter screen located beneath it for damage or contamination. The filter may be cleaned by removing it from the carburettor body and rinsing it in clean fuel, gently removing any stubborn contamination with a small soft-bristled brush. Renew the sealing washer beneath the head of the needle seat.

12   Unscrew and remove the main jet followed by the needle jet holder. When unscrewing any jet, a close fitting screwdriver must be used to prevent damage to the slot in the soft jet material. The needle jet should fall from its housing when the carburettor is tilted towards its fitted position. If the jet proves difficult to remove by this method then it must be carefully pushed out from the carburettor venturi side by using a small soft-wood block or something that will not damage the jet face.

13   A diaphragm air cut-off valve is fitted to each carburettor to richen automatically the mixture on over-run, thus preventing backfiring in the exhaust system. The valve is located on the side of the main body, and thus will require separation of the instruments if attention is required. The cut-off valve is enclosed by a cover held on the outside of the carburettor body by two screws. Unscrew the screws holding the cover in place against the pressure of the diaphragm spring, and then lift the cover away. Remove the spring, and carefully lift out the diaphragm. Note the small O-ring in the carburettor body mating face. Renew this O-ring and inspect the valve diaphragm for damage or deterioration. The spring should also be inspected for breakage.

14   Note that No 3 carburettor has an accelerator pump mounted on its underside; this should now be inspected in accordance with the following Section of this Chapter.

15   Before removing the pilot screw, screw it fully in until it seats gently counting and recording the number of turns required to do so. Failure to observe this precaution will make it necessary to reset the carburettor pilot screw settings on reassembly. Unscrew and remove the pilot screw followed by its spring. Tilt the carburettor to remove the plain washer and O-ring.

16   It is not recommended that the choke butterfly valve be removed as the valve itself is not subject to wear. If wear has occurred on the operating pivots, a new carburettor will be required as air will find its way along the pivot bearings resulting in a weak mixture.

17   Before the carburettors are reassembled, each should be cleaned out thoroughly, using compressed air. Avoid using a piece of rag since there is always risk of particles of lint obstructing the airways or jet orifices. Never use a piece of wire or any pointed metal object to clear a blocked jet. it is only too easy to enlarge a jet under these circumstances and increase the rate of petrol consumption. If an air line is not available, a blast of air from a tyre pump will usually suffice.

18   Reassemble each carburettor, using the reversed dismantling procedure. Work must be carried out in absolute cleanliness. If possible use new gaskets; O-rings can be reused if there is no doubt as to their condition. When replacing the piston, note that it can be refitted in only one position. The groove in the piston side must locate with the projection in the main body. Assemble the carburettors in their pairs on the air box halves and then screw the halves together. Reconnect the choke operating rod and the throttle control rods. Remember to use new split-pins and to refit the plain washers beneath the pins before spreading the pin legs. Check the choke mechanism for correct operation.

19   Do not use excessive force when reassembling a carburettor since it is easy to shear a jet or some of the smaller screws. Furthermore, the carburettors are die cast in a zinc-based alloy which itself does not have a high tensile strength.

9.5a Remove the gauze screen from inside the air box half ...

9.5b ... detach the chrome stay plate ...

9.5c ... and carefully separate the two carburettors noting the fuel joint pipes

9.6 Loosen the manifold clamp retaining screw

9.8a Carefully withdraw the vacuum piston ...

9.8b ... and remove the plug complete with O-ring

9.8c Remove the needle from the piston ...

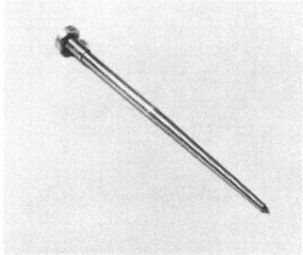

9.8d ... and inspect the needle for wear or damage

9.9a Remove the screw and plate from the carburettor body ...

9.9b ... to reveal the air jet located beneath it

9.10a Lift the float chamber from position ...

9.10b ... displace the float pivot pin ...

9.10c ... and remove the twin float assembly

9.11 Remove the float needle seat and filter screen assembly

9.12a Unscrew and remove the main jet ...

9.12b ... followed by the needle jet holder

9.12c ... and the needle jet

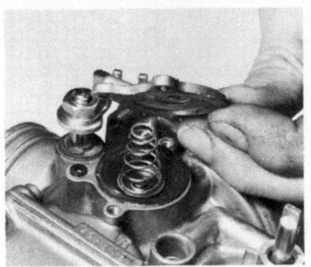

9.13a Remove the air cut-off valve cover ...

9.13b ... to allow removal of the diaphragm

9.18a Locate the groove in the piston side with the projection in the carburettor body

9.18b Note the location of the spring between the throttle links

9.18c Carefully push the air box halves together ...

9.18d ... and fit and tighten the securing screws, noting the location of the choke cable guide

9.18e Reconnect the choke operating rod and lever

9.18f Reconnect the throttle control rods as shown (arrow indicates the throttle stop screw)

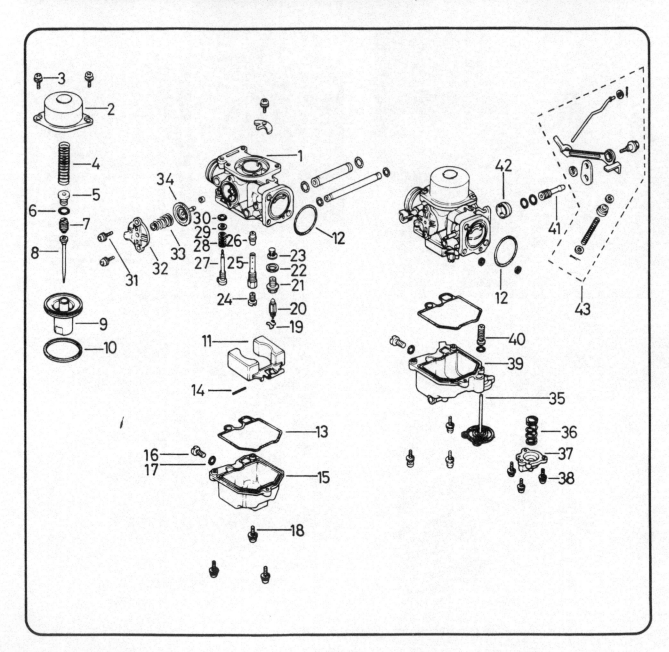

Fig. 3.5 Carburettor

1	Carburettor body	24	Main jet
2	Vacuum chamber	25	Needle jet holder
3	Screw – 2 off	26	Needle jet
4	Spring	27	Pilot screw
5	Plug	28	Spring
6	O-ring	29	Washer
7	Needle retaining	30	O-ring
	screw	31	Screw – 2 off
8	Piston needle	32	Diaphragm cover
9	Vacuum piston	33	Diaphragm spring
10	Seal	34	Diaphragm air cut-
11	Float		off valve
12	O-ring	35	Diaphragm/operating
13	Sealing ring		rod
14	Pivot pin	36	Spring
15	Float chamber	37	Diaphragm cover
16	Drain screw	38	Screw – 3 off
17	Sealing washer	39	Float chamber
18	Screw – 3 off		(No. 3 carburettor)
19	Float needle clip	40	Gaiter
20	Float needle	41	Fuel feed union
21	Needle seat	42	Choke dust cap
22	Sealing washer	43	Accelerator pump
23	Filter screen		operating assembly

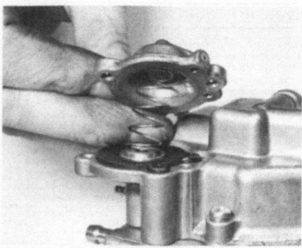

10.3a Remove the accelerator pump cover ...

10 Accelerator pump: examination and adjustment

1   The accelerator pump is mounted on the underside of No 3 carburettor. Its purpose is to richen the mixture during acceleration, thus allowing the carburettor to be jetted for a weaker overall mixture to meet EPA emission requirements in the USA.
2   The pump is operated by a spring rod connected to the throttle operating linkage. Once actuated by throttle opening, the pump feeds fuel into No 3 carburettor only.
3   In the event of pump failure, the pump can be dismantled by unscrewing the three pump cover retaining screws and removing the cover and spring. This will expose the pump diaphragm and operating rod assembly. Remove the diaphragm and inspect it for signs of damage or deterioration. Check the rod is not bent and the spring is not broken. Check also the condition of the small gaiter at the carburettor body end of the operating rod and renew it if it is found to be split or perished. Note that it will be necessary to remove the float chamber to allow the gaiter to be withdrawn from its location in the carburettor body. Little should go wrong with the pump apart from a cracked or perished diaphragm, but if the pump system is completely dry it may require priming to expel air.
4   Check the pump adjustment by measuring the clearance between the stop on the carburettor body and the adjuster arm on the pump operating linkage. The correct clearance should be 10 mm (0.4 in). If necessary, alter the clearance by bending the adjuster arm.

10.3b ... to enable the pump diaphragm and rod to be lifted out of the carburettor body

11 Carburettors: idle speed adjustment

1   Adjustment of tick-over on most motorcycles is made by careful adjustment of the pilot screws in conjunction with the throttle stop screw(s). On the Honda GL1100 the pilot screw on each carburettor is adjusted at the factory and should not be altered unless new pilot screws are being fitted. Tick-over adjustment is therefore limited to alteration of the throttle stop screw located immediately below the throttle operating pulley.
2   The adjustment of engine speed should be made only when the engine has reached normal working temperature. The correct tick-over speed is 950 ± 100 rpm.

10.3c Inspect the pump rod gaiter

12 Carburettors: pilot screw adjustment

1   Before attempting synchronisation of the carburettors, ensure that the pilot screw adjustment is correct. It will be found that on USA market models, the pilot screws have a limiter cap cemented to them, thus preventing any adjustment other than to weaken the mixture.

2   Note that it is not necessary to carry out this adjustment unless the carburettor has been dismantled and the pilot screw disturbed. Fit the pilot screw and screw it in until it is felt to come into contact with its seat. Do not tighten the screw against the seat otherwise damage to the seat will occur. Unscrew the pilot screw 1¼ turns to obtain the initial setting.

3   Obtain and fit a tachometer with graduations of 50 rpm or less that will accurately indicate a 50 rpm change. Start the engine and allow it to reach normal operating temperature. Adjust the engine idle speed to 950 ± 100 rpm by turning the throttle stop screw.

4   Turn each pilot screw out by ½ turn. If the engine speed increases by 50 rpm or more, each pilot screw should be turned out a further ½ turn and the operation repeated if necessary until the engine speed drops by 50 rpm or less. Reset the engine idle speed to 950 ± 100 rpm by turning the throttle stop screw.

5   Move to No 1 carburettor and turn the pilot screw in until the engine speed drops by 50 rpm. Note the position of the pilot screw head and turn the pilot screw one complete turn out from that position. Reset the engine idle speed to 950 ± 100 rpm by turning the throttle stop screw. Repeat the procedure given for No 1 carburettor in this paragraph on No 2, 3 and 4 carburettors. The carburettors should now be synchronised as described in the following Section.

12.1 The pilot screw (arrowed) is located at the base of the carburettor

13 Carburettors: synchronisation

1   For the best possible performance it is imperative that the carburettors are working in perfect harmony with each other. If the carburettors are not synchronised, not only will one cylinder be doing less work, at any given throttle opening, but it will also in effect have to be carried by the other cylinders. This will reduce performance accordingly.

2   For synchronisation, it is essential to use a vacuum gauge set consisting of four separate dial gauges, one of which is connected to each carburettor by means of a special adaptor tube. The adaptor pipe screws into the outside lower end of each inlet manifold, the orifice of which is normally blocked off by a cross-head screw plug. A suitable vacuum gauge set may be purchased from a Honda Service Agent, or from one of the

many suppliers who advertise regularly in the motorcycle press.

3   Bear in mind that this equipment is not cheap, and unless the machine is regarded as a long-term purchase, or it is envisaged that similar multi-cylinder motorcycles are likely to follow it, it may be better to allow a Honda dealer to carry out the work. The cost can be reduced considerably if a vacuum gauge set is purchased jointly by a number of owners. As it will be used fairly infrequently this is probably a sound approach.

4   Place the vacuum gauge set on the machine so that the dials can be easily read. The usual position is between the handlebars. Remove the four blanking plugs from the manifolds and fit the adaptor pieces. Connect the gauges as follows for ease of observation.

No. 1 cylinder	right-hand gauge
No. 2 cylinder	inner right-hand gauge
No. 3 cylinder	inner left-hand gauge
No. 4 cylinder	left-hand gauge

Start the engine allowing it to warm up, and set the speed to 950 ± 100 rpm by turning the throttle stop screw. Note that if the readings on the gauges fluctuate wildly, it is likely that the gauges require heavier damping. Refer to the gauge manufacturer's instructions on setting up procedures.

5   Note the reading on each gauge. Select the gauge which shows the reading nearest to midway between the highest and lowest readings of the four gauges and use this reading as a datum upon which to set the other three carburettors.

6   Pressure between single carburettors in the same pair may be varied by means of the adjustment screw and locknut located between the instruments. Moving this screw in one direction will raise the vacuum on one carburettor and reduce it on the other. Once the single carburettors in the same pair have been set to give as close a reading as possible on their corresponding gauges, the adjustment screw on the rearmost face of No 4 carburettor may be used to regulate the pressures between the two pairs of carburettors.

7   With all four gauges reading as close to the selected datum as possible, set the engine idle speed to 950 ± 100 rpm and stop the engine. A difference in vacuum between the carburettors of less than 50 mm Hg (2.0 in Hg) should be aimed at to ensure smooth running. Remove the adaptors, pipes and gauges and refit the manifold blanking plugs.

14 Carburettors: settings

1   Some of the carburettor settings, such as the sizes of the needle jets, main jets and needles etc, are determined by the manufacturer. Under normal circumstances it is unlikely that these settings will require modification, even though there is provision made. If a change appears necessary it can often be attributed to a developing engine fault.

2   It is recommended that changes in carburettor specifications are not made as the manufacturer will have spent considerable research time in assessing performance of the engine and will have come to a satisfactory conclusion. The only condition where settings may require alteration are when the machine is to be used at very high altitude when touring the more mountainous parts of the globe. In these cases it is probable that the main jet sizes will require reducing to correspond with the change in atmosphere. Consult a Honda specialist before any such alterations are made.

15 Carburettors: adjusting float level

1   If problems are encountered with fuel overflowing from the float chambers, which cannot be traced to the float/needle assembly, or if consistent fuel starvation is encountered, the fault will probably lie in maladjustment of the float level. It will be necessary to remove the float chamber bowl from each

carburettor to check the float level.

2   If the float level is correct the distance between the uppermost edge of the floats and the flange of the mixing chamber body will be 15.5 mm (0.61 in).

3   Adjustments are made by bending the float assembly tang (tongue) which engages with the float tip, in the direction required (see accompanying diagram).

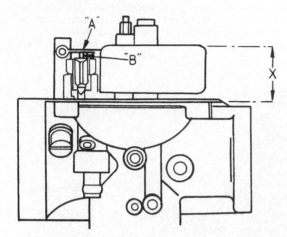

Fig. 3.6 Checking float level

A   Float tongue   X   Float height
B   Float valve

16 Air filter element: renewal

1   To remove the air filter element from the machine, open the dummy fuel tank compartment covers and lift out the tool tray. Unscrew and remove the single wingnut and lift out the air filter cover along with its seal. Remove the air filter element and seal from the filter housing, discard the element and inspect the seal for damage and deterioration before refitting a serviceable item. Fit a new filter element, the filter cover and seal, refit and tighten the wingnut and refit the tool tray before closing the compartment covers.

2   Do not run the engine with the element removed as the weak mixture caused may result in engine overheating and damage to the cylinders and pistons. A weak mixture can also result if the rubber sealing rings on the element are perished or omitted.

17 Exhaust system: removal and refitting

1   Unlike the 2-stroke engine, the exhaust gases of a 4-stroke engine are usually not of an oily nature. The Honda GL 1100 silencer is therefore not fitted with detachable baffles. If any fault develops in the silencer, the complete system half should be renewed. Removal of the exhaust system may be more easily achieved with the aid of an assistant and should be carried out as follows.

2   Remove the two flange nuts from each of the four exhaust pipe to cylinder head connections. Remove the two bolts from the balancer pipe clamp and check that the clamp is free. Loosen the nut and bolt retaining each silencer to each pillion footrest mounting bracket and remove each nut. With a person each side of the machine supporting the exhaust assembly, withdraw the mounting bolt from each pillion footrest bracket and carefully lower the complete assembly away from its

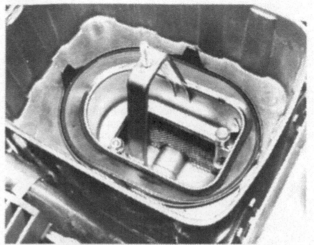

16.1a Inspect the filter seal for damage and deterioration ...

16.1b ... before fitting a new filter element ...

16.1c ... followed by the filter cover and wingnut

mounting points. Once clear, the balancer pipe may be separated at its connection and the two separate exhaust pipes lifted clear of the machine. In no circumstances should the exhaust assembly be allowed to hang unsupported from its cylinder head mounting points because the weight of the system will place an unacceptable strain on the cylinder head studs.

3   As with removal, refitting of the exhaust system requires the aid of an assistant. Fit a new gasket into each exhaust port, distorting them very slightly if necessary so that they become oval and remain in the ports. With a person either side of the machine, push the two halves of the system together at the balancer pipe connection. The complete assembly may then be lifted up to locate on the cylinder head studs and align with the rear mounting points. Refit the bolt and nut to each rear

mounting and the two flange nuts to each of the four cylinder head manifold connections. Tighten the manifold nuts, followed by the rear mounting nuts and the two bolts retaining the balancer pipe clamp. Take care to tighten the manifold nuts evenly to avoid distortion of the pipe flanges. Note the torque figures given in the Specifications of this Chapter.

4   Do not run the machine with the exhaust baffles removed, or with a quite different type of silencer fitted. The standard production silencers have been designed to give the best possible performance, whilst subduing the exhaust note to an acceptable level. Although a modified exhaust system, or one without baffles, may give the illusion of greater speed as a result of the changed exhaust note, the chances are that performance will have suffered accordingly.

17.2 Two bolts retain the balancer pipe clamp in position

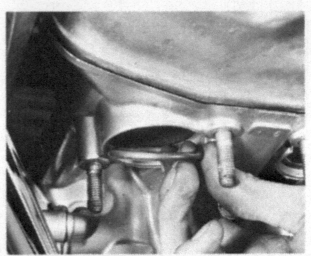

17.3 Fit a new gasket into each exhaust port

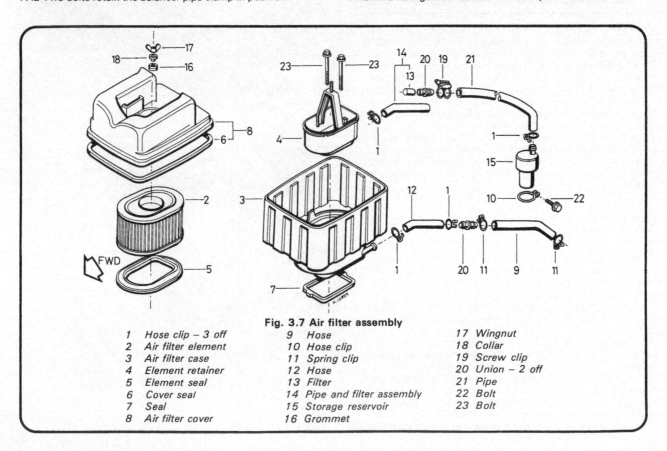

Fig. 3.7 Air filter assembly

1   Hose clip – 3 off
2   Air filter element
3   Air filter case
4   Element retainer
5   Element seal
6   Cover seal
7   Seal
8   Air filter cover
9   Hose
10 Hose clip
11 Spring clip
12 Hose
13 Filter
14 Pipe and filter assembly
15 Storage reservoir
16 Grommet
17 Wingnut
18 Collar
19 Screw clip
20 Union – 2 off
21 Pipe
22 Bolt
23 Bolt

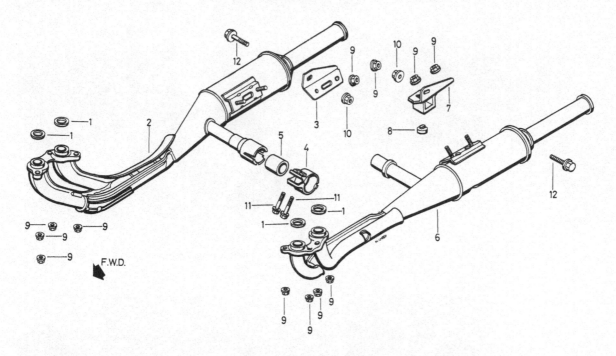

Fig. 3.8 Exhaust system

1 Exhaust port gasket – 4 off	5 Balancer pipe seal	9 Flange nut – 12 off
2 Right-hand exhaust pipe	6 Left-hand exhaust pipe	10 Flange nut – 2 off
3 Right-hand retaining bracket	7 Left-hand retaining bracket	11 Bolt – 2 off
4 Balancer pipe clamp	8 Main stand rubber stopper	12 Bolt – 2 off

18 Oil pumps: removal, examination and fitting

1   Unless lubrication failure occurs, the oil pumps should be checked whenever an engine overhaul is undertaken. The main oil pump can be removed from the front of the engine whilst the engine is still in the frame, after following the directions for radiator removal and water pump removal in Sections 5 and 8 respectively of Chapter 2. The pump retaining screws can then be removed and the pump pulled from position, together with the drive shaft. Removal of the clutch scavenge pump can only take place after the engine has been removed from the frame and the clutch itself has been removed. Refer to Chapter 1 for engine removal and dismantling.

2   Dismantle the two pumps separately, cleaning them thoroughly in petrol and allowing them to dry before inspection is carried out. Check the castings for signs of cracking or fracture and inspect the inner wall of the main bodies for scoring. Using a feeler gauge, check the side clearances and radial clearance of each pump and rotors, which should fall within the specifications as outlined at the beginning of the Chapter. Rotors must be renewed as a set and if the pump body is worn, the complete assembly must be renewed.

3   Examine the rotors and pump bodies for signs of scoring chipping or other surface damage, which will occur if metallic particles find their way into the oil pump assembly. Renewal of the affected parts is the only remedy under these circumstances.

4   Reassemble the pumps in absolutely clean conditions. Even a small particle of grit or metal may damage the rotors. Remember that the punch-marked faces of the rotors must face away from the main pump body. Before fitting the rotors into the pump body, lubricate them thoroughly with clean engine oil.

5   Refit the pumps in the engine by reversing the removal procedure. Ensure that all O-rings are correctly fitted and not damaged or deteriorated and that the hollow dowels and paper gasket of the main pump are correctly positioned.

19 Oil filter: renewing the element

1   The oil filter is contained within a separate housing, bolted to the front engine cover by a single hollow bolt containing the pressure release valve. Access to the element is made by unscrewing the centre bolt, which will bring with it the housing and element. Before removing the housing, place a receptacle beneath the engine to catch the engine oil contained in the filter chamber.

2   When renewing the filter element, it is wise to renew the housing O-ring at the same time. An O-ring is usually provided with a filter.

3   The filter by-pass valve (pressure release valve), comprising a plunger and spring, is situated in the bore of the centre bolt. It is recommended that the by-pass valve be checked for free operation at regular intervals. The spring and plunger are retained by a pin across the centre bolt. Tap the pin out, to allow removal of the plunger and spring for cleaning.

4   When refitting the housing to the engine, ensure it is aligned so that the tabs on its edge fit one either side of the boss on the water pump cover. Tighten the housing bolt to a torque of 2.7 – 3.3 kgf m (20 – 24 lbf ft).

5   Never run the engine without the filter element or increase the period between the recommended oil changes or oil filter changes. Change the engine oil and filter every 8000 miles or every year, whichever is sooner.

18.2a Remove the main pump cover...

18.2b ... to gain access to the rotors

18.2c Note the drive pin and thrustwasher

18.2d Measure rotor to body clearance ...

18.2e ... and inner rotor to outer rotor clearance

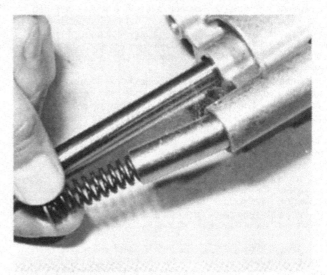

18.2f Spring and seat of the main pump pressure release valve ...

18.2g ... are retained by a split-pin

18.3a Examine the pump rotors for the kind of scoring shown

18.3b Measure the rotor to body side clearance

18.4a Note the position of the scavenge pump body locating pins ...

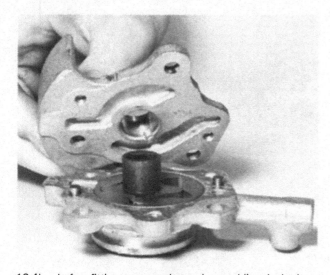

18.4b ... before fitting a new gasket and assembling the body halves

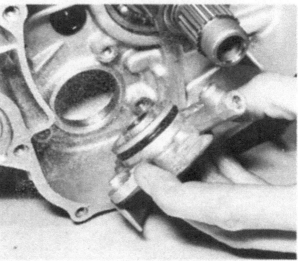

18.5a Fit a new O-ring ot the scavenge pump boss ...

18.5b ... and secure the pump body to the engine casing

18.5c Note the dowel and O-rings fitted to the main pump body

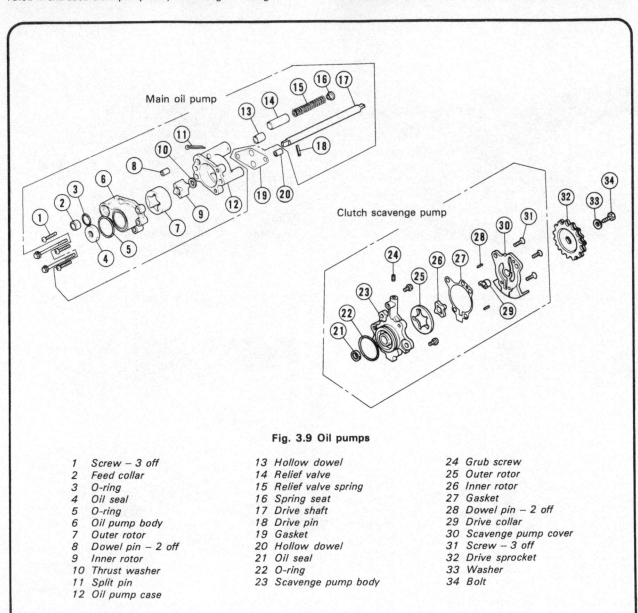

Fig. 3.9 Oil pumps

1 Screw – 3 off	13 Hollow dowel	24 Grub screw
2 Feed collar	14 Relief valve	25 Outer rotor
3 O-ring	15 Relief valve spring	26 Inner rotor
4 Oil seal	16 Spring seat	27 Gasket
5 O-ring	17 Drive shaft	28 Dowel pin – 2 off
6 Oil pump body	18 Drive pin	29 Drive collar
7 Outer rotor	19 Gasket	30 Scavenge pump cover
8 Dowel pin – 2 off	20 Hollow dowel	31 Screw – 3 off
9 Inner rotor	21 Oil seal	32 Drive sprocket
10 Thrust washer	22 O-ring	33 Washer
11 Split pin	23 Scavenge pump body	34 Bolt
12 Oil pump case		

20 Oil pressure warning light

1   An oil pressure warning light is incorporated in the lubrication system to give immediate warning of excessively low oil pressure.

2   The oil pressure switch is screwed into the top of the crankcase, below No 3 carburettor. The switch is connected to a warning light on the handlebar mounted lighting console. The light should be on when the ignition is switched on but will usually go out almost as soon as the engine is started.

3   Should the oil pressure warning light come on whilst the machine is being ridden, the engine should be switched off immediately, otherwise there is a risk of severe engine damage due to lubrication failure. The fault must be located and rectified before the engine is re-started and run, even for a brief time. Machines fitted with plain shell bearings rely on high oil pressure to maintain a thin oil film between the bearing surfaces. Failure of oil pressure will cause the working surfaces to come into direct contact, causing overheating and eventual engine seizure.

20.2 The oil pressure switch is screwed into the top of the crankcase

21 Fault diagnosis: fuel system and lubrication

Symptom	Cause	Remedy
Engine gradually fades and stops	Fuel starvation	Check vent hole in filler cap Check for sediment in fuel filter and renew if necessary Remove float chamber and check for sediment in both chamber and float needle seat filter screen
Engine runs badly. Black smoke from exhausts	Carburettor flooding	Dismantle and clean carburettor. Check for punctured float or sticking float needle Check fuel level in float chamber
	Air filter blocked	Renew filter
Engine lacks response and overheats	Weak mixture Air cleaner disconnected, seals or hoses split Modified silencer has upset carburation	Check for partial block in carburettors Reconnect or renew as necessary Replace with original design
Oil pressure warning light comes on	Lubrication system failure	Stop engine immediately. Trace and rectify fault before re-starting
Engine gets noisy	Failure to change engine oil when recommended	Drain off old oil and refill with new oil of correct grade. Renew oil filter element.

Chapter 4 Ignition system

Contents

Specifications

Ignition system
Type .... Capacitor discharge ignition (CDI)

Ignition timing
Initial .... 13°BTDC @ 950 ± 100 rpm
Fully advanced .... 38.5° BTDC @ 2950 rpm

Ignition coils
No off .... Two, twin coils
Make .... Toyo Denso

Sparking plug
Gap .... 0.6 – 0.7 mm (0.024 – 0.028 in)
Firing order .... 1 – 3 – 2 – 4

Type:	USA	UK
Standard	NGK/ND	NGK/ND
	D8EA/X24 ES-U	DR8ES-L/X24ESR-U
Cold climate (below 5°C, 41°F)	NGK/ND	NGK/ND
	D7EA/X22ES-U	DR7ES/X22ESR-U
Extended high speed riding	NGK/ND	NGK/ND
	D9EA/X27ES-U	DR8ES/X27ESR-U

1 General description

The Honda GL 1100 model makes use of a capacitor discharge ignition (CDI) system, in which the ignition timing and triggering is controlled electronically. As a result, the spark at the plugs is more powerful and is accurately timed. Because there are no mechanical contact breakers, wear does not take place, and thus the ignition timing will remain accurate unless disturbed, or in the rare event of component failure.

Like most 4-cylinder machines, the Honda GL 1100 employs what is effectively two ignition systems, each controlling two cylinders. This means that each time a spark is generated in one of the ignition coils, it is fed to two of the sparking plugs. Combustion will only take place in the cylinder which is under compression, leaving one spark wasted. This arrangement is known as the 'spare spark' system. It is important to remember this when dealing with ignition system faults, as these are most likely to be confined to one half of the system. The system can therefore be divided into that which relates to cylinders 1 and 2, and the corresponding equipment relating to cylinders 3 and 4.

The system is triggered by a reluctor mounted on the crankshaft ends which rotates past two pickup coils or pulsers.

The reluctor is analogous to the contact breaker cam in a conventional system. As the peak of the reluctor passes the pulser coil, it generates a small signal current. This current triggers the appropriate circuit in the ignition amplifier, or spark unit. The high voltage pulse from the amplifier feeds to the relevant coil, where its passage through the coil's primary windings induces the required high tension (HT) spark in the secondary windings. The reluctor continues to rotate, causing the same chain of events to take place in the remaining circuit.

For the GL 1100 model, Honda have chosen to retain a mechanical automatic timing unit (ATU) rather than opt for full electronic advance. As the engine speed rises, small weights on the ATU are thrown outwards against spring pressure. This movement is translated to the reluctor, which advances in relation to the crankshaft, thus providing an advanced spark for high speed running.

Additional advance control is provided by a vacuum advance unit mounted on the pulser generator housing. Any changes in manifold depression are transmitted from an air inlet adaptor on No 3 carburettor via a small bore hose to the sealed side of the advance unit diaphragm casing. This in turn causes the diaphragm to move which transmits movement to the baseplate of the pulser coils via an operating arm, thus advancing the spark in relation to throttle opening.

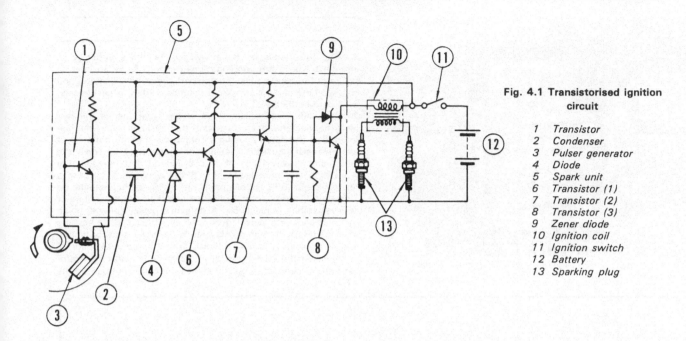

Fig. 4.1 Transistorised ignition circuit

1   Transistor
2   Condenser
3   Pulser generator
4   Diode
5   Spark unit
6   Transistor (1)
7   Transistor (2)
8   Transistor (3)
9   Zener diode
10 Ignition coil
11 Ignition switch
12 Battery
13 Sparking plug

2   Electronic ignition system: testing

1   The CDI system, as mentioned previously, can be divided into two separate systems for most fault diagnosis purposes. Total failure of the entire system is unlikely, and trouble is usually confined to 1 and 2 cylinders or to 3 and 4 cylinders. In this case, it can be assumed that the system is functional up to the amplifier, or spark unit.

2   A preliminary check may be made as follows. Remove the sparking plugs from cylinders 1 and 3, remembering that these each represent one sub-system. With the sparking plugs reconnected to their respective HT leads, place each plug on the cylinder head casting.

3   At this point, Honda recommend that the cover be removed from the pulser generator housing to enable a screwdriver with an insulated handle to bridge the gap between the rotor and the metal core of one of the pulser coils. With the ignition switched on a spark should occur at the appropriate plug, indicating that the ignition circuitry is sound for that sub-system. It is then recommended that the process be repeated for the other sub-system by bridging the gap between the remaining pulser coil and the rotor. No spark at either one of the plugs will indicate a defect in the system. During this test ensure that the sparking plugs remain earthed at all times. Failure to do this may overload the system causing a malfunction in the CDI unit.

4   It should be noted that removal of the pulser generator cover and access to the rotor and coils within is not possible unless the rear wheel and fuel tank assemblies are removed beforehand. For this reason above, it is recommended that the following test is carried out as an alternative to that given above.

5   With the machine positioned on its main stand and with the gearchange lever selected to neutral, switch the ignition on and press the starter button. Directly the engine begins to turn, a spark should be observed at each of the sparking plugs. No spark at either one of the plugs will indicate a defect in the system.

6   If no spark at all is found, attention should be turned to the ignition switch. Check that the switch operates properly, and that power is being fed to the ignition amplifier (spark unit). Ensure that the kill switch is at the 'run' position and has not shorted. If all seems well, and the remaining wiring connections are sound, it can be assumed that the amplifier is in need of

attention. As it is unlikely that both pulser coils, both ignition coils or all four sparking plugs will have failed simultaneously, they can be ignored for the time being.

7   If one of the plugs failed to produce a spark in the above test, the system can be assumed to be sound up to the amplifier unit. The latter should be checked, followed by the pulser coil(s) and ignition coil(s).

3   Ignition amplifier (spark unit): testing

1   If preliminary checks have indicated a possible fault in the amplifier unit it should be tested as described below. Note that a dc voltmeter, or a multimeter set on the appropriate scale, will be required for the test. It is assumed that owners with access to this equipment will be conversant with its use. If this is not the case, the work should be entrusted to a Honda Service Agent or an Auto-electrician.

Test A

2   Remove the dummy fuel tank and locate the 6-pole block connector, which is the uppermost of a batch of three located just forward of the spark unit itself. Connect a multimeter, set on 0 – 50 volts dc, as follows. Attach the positive (+) probe to the blue/yellow wire terminal (3 – 4 cylinders) and the negative (-) probe to earth. Turn the ignition switch to the 'On' position. This should result in the meter needle fluctuating between 12V and 0V if the unit is working correctly. Repeat this test, connecting the positive probe to the yellow wire terminal (1-2 cylinders) and the negative probe to earth.

Test B

3   Remove the left-hand sidepanel and disconnect the leads to the pulser generator at the innermost of the two block connectors located just forward of the battery. Attach the positive probe of the multimeter to the white/blue terminal (3-4 cylinders). Touch the negative probe of the multimeter to earth and note the reading on the scale of the meter; this should be between 12V and 0V if the spark unit is serviceable. Repeat this test, attaching the positive probe to the white wire terminal (1-2 cylinders).

4   If the unit proves to be defective, then it must be renewed since there is no practicable method of repair.

3.2 Locate the ignition amplifier (spark unit) and its block connector

4   Pulser coils: testing

1   The pulser coils can fail in two possible ways; either by an internal break in the windings (open circuit) or by an internal failure of the winding insulation (short circuit). A pulser in good condition will show a specific resistance reading when checked as described below.

2   Disconnect the pulser leads at the innermost of the two block connectors, located beneath the left-hand sidepanel just forward of the battery. Using an ohmmeter or a multimeter set on the resistance scle, measure the resistance between the white and yellow leads (1 − 2 cylinders) and between the white/blue and blue leads (3 − 4 cylinders). In each case a resistance of 530 ± 50 ohms at 20°C (68°F) should be indicated. Open or short circuits will be indicated by infinite or zero resistance respectively. A defective pulser coil will mean that the complete pulser generator assembly and its housing must be returned to an official Honda Service Agent for testing and, if necessary, renewal. Check the letter code (A, B or C) stamped on the housing of any replacement pulser generator assembly and ensure that it matches the code stamped on the ATU.

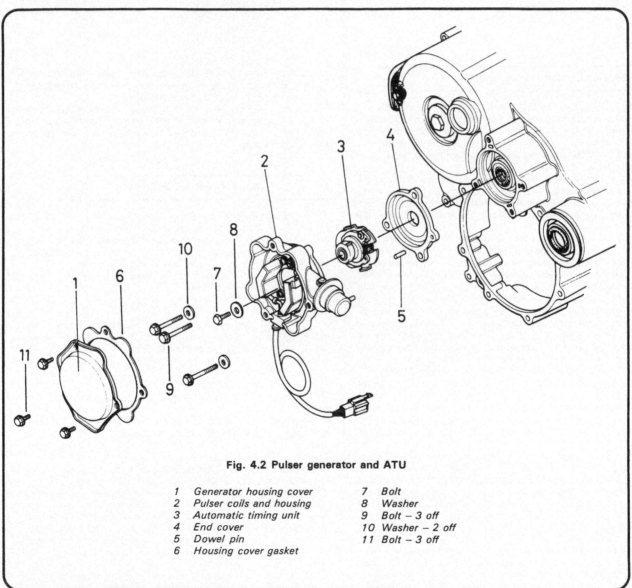

Fig. 4.2 Pulser generator and ATU

1	Generator housing cover	7	Bolt
2	Pulser coils and housing	8	Washer
3	Automatic timing unit	9	Bolt – 3 off
4	End cover	10	Washer – 2 off
5	Dowel pin	11	Bolt – 3 off
6	Housing cover gasket		

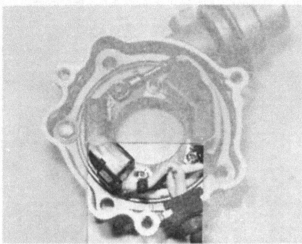

4.2 Check the letter code (B) stamped on the pulser generator housing

5 Automatic timing unit (ATU): examination

1 The automatic timing unit rarely gives rise to problems unless of considerable age or if neglect has allowed the pivots to become corroded. As CDI ignition is fitted attention to this area is likely to be far from frequent. It is worth checking the unit whenever the engine is removed from the frame as this is the only time access can be gained to this component. Full details of unit removal are given in Section 12 of Chapter 1.

2 The unit comprises spring loaded balance weights, which move outward against the spring tension as centrifugal force increases. The balance weights must move freely on their pivots and be rust-free. The tension springs must also be in good condition. Keep the pivots lubricated and make sure the balance weights move easily, without binding. Most problems arise as a result of condensation, within the engine, which causes the unit to rust and balance weight movement to be restricted.

3 The automatic timing unit mechanism is fixed in relation to the crankshaft extension by means of a locating pin, which must also be checked for wear. In consequence, the mechanism cannot be fitted in anything other than the correct position. This ensures accuracy of ignition timing to within close limits.

4 If any doubt exists as to the condition of the unit, seek the advice of an official Honda Service Agent who will be able to give an expert opinion and supply a replacement unit if considered necessary. It must be noted that there is no method of determining whether or not the unit is functioning correctly once it is fitted to the engine and the engine fitted in the frame. It is, therefore, essential that the unit is fully serviceable when fitted.

5 If the automatic timing unit has to be renewed, then it will be necessary to separate the old unit from the rotor by pulling the two components apart. Note that the tooth on the rotor must align with the cutout in the edge of the automatic timing unit when assembling the two components. Note that the letter code (A, B or C) stamped on the ATU must match that stamped on the pulser generator housing.

6 Vacuum advance unit: testing

1 The vacuum advance unit forms part of the pulser generator assembly and comprises a diaphragm to which is connected an operating arm which in turn is connected to a point on the baseplate of the pulser coils.

2 The purpose of the unit is to provide additional advance control in response to engine vacuum, and thus load, whilst working in configuration with the ATU to provide a better controlled and more responsive spark. It achieves this by means of a small bore hose running from the air inlet adaptor on the float chamber of No 3 carburettor to an adaptor on the sealed side of the advance unit diaphragm casing. Any changes in manifold depression cause the diaphragm to move, which in turn moves the pulser coil baseplate through a small arc by means of the operating arm.

3 If the unit is suspected of being defective, check the hose for damage or deterioration or any signs of leakage at its end connection before progressing to the following test sequence. Disconnect the hose from the carburettor adaptor and attach a vacuum pump and gauge to its end. The unit is operating correctly if the following readings are obtained.

All models
Operation of unit starts at 40 mm Hg
All 1980 models and 1981 (UK) models
Operation of unit ceases at 100 mm Hg
1981 (US) models
Operation of unit ceases at 80 mm Hg

If these readings are not obtained, the unit is defective and the complete pulser generator assembly must be renewed.

4 If the vacuum pump and gauge are not obtainable, the machine may be returned to an official Honda Service Agent who will be able to carry out the check. As an alternative, suck through the end of the hose whilst observing the movement of the baseplate.

5 Because of the difficulty of removing the housing cover with the engine in situ in order to observe the movement of the pulser coils and baseplate, it is well worth checking the unit for full and smooth operation whenever the engine is removed from the frame.

7 Ignition coils: location and testing

1 Two separate ignition coils are fitted, each of which supplies a different pair of cylinders. The coils are mounted within the dummy fuel tank, one either side of the frame top tube.

2 If a weak spark, poor starting or misfiring causes the performance of the coils to be suspected, they should be tested by a Honda Service Agent or an auto-electrician who will have the appropriate test equipment.

3 It is unlikely that the coils will fail simultaneously. If intermittent firing occurs on one pair of cylinders the coils may be swopped over by interchanging the low tension terminal leads and the HT leads. If the fault then moves from one pair of cylinders to the other, it can be taken that the coil is faulty.

4 The coils are sealed units and therefore if a failure occurs, repair is impracticable. The faulty item should be replaced by a new component.

7.1 The HT coils are mounted within the dummy fuel tank

7.3 The HT leads may be unscrewed from the coil connections

8 Ignition timing

1   Ignition timing is set accurately at the manufacturing stage and fixed in position. No provision is made for timing adjustment. Static timing is set by means of the ATU being aligned with the crankshaft extension locating pin and the pulser coils with their housing being aligned with the engine casing locating pin.
2   The ignition timing may be checked dynamically by removing the cap from the timing mark hole in the engine crankcase and inserting in its place the inspection cap, Honda tool no 07999-3710001. This tool incorporates a sighting glass which is crossed by lines corresponding to those on the edge of the crankcase hole and represents the only accurate means of aligning the timing marks.
3   Connect a good quality stroboscopic lamp to the HT lead of No 1 or No 2 cylinder and start the engine, allowing it to idle at 950 ± 100 rpm. The F1 timing mark on the flywheel should align with the sighting marks on the glass of the tool. Repeat the procedures with the lamp connected to No 3 or No 4 cylinder HT lead. The F2 timing mark on the flywheel should now align with the sighting marks of the tool.
4   If it is found that the timing marks do not align as described above, refer to the information contained within this Chapter for test on the system components. On completion of the timing check, disconnect the stroboscopic lamp and refit and tighten the cap in the timing mark hole.

9 Sparking plugs: checking, cleaning and resetting the gaps

1   The Honda GL 1100 is fitted with NGK or ND sparking plugs of various types. These are described in the specifications Section at the beginning of the Chapter. Note that in certain

operating conditions, a change of plug grade may be required, though the standard grade will prove adequate in most cases.
2   To maintain optimum engine performance, Honda recommend that the plugs are changed every 4000 miles (6400 km). The points gap of the plugs should be checked at frequent intervals during this period. To reset the gap, bend the outer electrode closer to or further away from the central electrode until the gap is within the range 0.6 – 0.7 mm (0.024 – 0.28 in) as measured with a feeler gauge. The gap is usually set to the lower figure to allow for erosion of the electrodes. Never bend the centre electrode or the insulator will crack, causing engine damage if the particles fall into the cylinder whilst the engine is running.
3   With some experience, the condition of the sparking plug electrodes and insulator can be used as a reliable guide to engine operating conditions.

4   Always carry a spare sparking plug of the recommended grade. In the rare event of plug failure, this will allow immediate replacement and prevent nuisance and the strain on the engine when attempting to continue on three cylinders.
5   Whenever the sparking plugs are removed, take the opportunity to clear out the drain channels which pass from the plug wells to the underside of the cylinder heads. If these become blocked and the machine is used in heavy rain the wells will fill up causing shorting in the suppressor caps and complete ignition failure.
6   If the threads in the cylinder head strip as a result of over tightening the sparking plugs, it is possible to reclaim the head by means of a Helical thread insert. This is a cheap and convenient method of replacing the threads; most motorcycle dealers operate a service of this nature at an economic price.
7   Make sure the plug insulating caps are a good fit and have their rubber seals. They should be kept clean to prevent leakage and tracking of the H/T current. These caps contain the suppressors that eliminate both radio and TV interference.

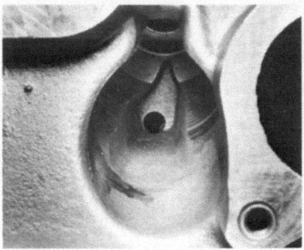

9.5 The drain channels in the sparking plug wells must be kept clear

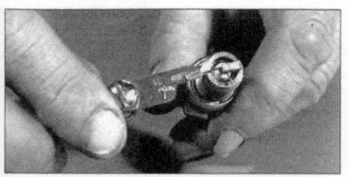

Electrode gap check - use a wire type gauge for best results

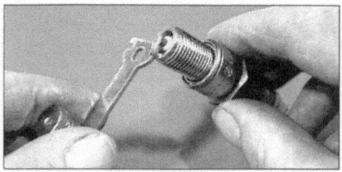

Electrode gap adjustment - bend the side electrode using the correct tool

Normal condition - A brown, tan or grey firing end indicates that the engine is in good condition and that the plug type is correct

Ash deposits - Light brown deposits encrusted on the electrodes and insulator, leading to misfire and hesitation. Caused by excessive amounts of oil in the combustion chamber or poor quality fuel/oil

Carbon fouling - Dry, black sooty deposits leading to misfire and weak spark. Caused by an over-rich fuel/air mixture, faulty choke operation or blocked air filter

Oil fouling - Wet oily deposits leading to misfire and weak spark. Caused by oil leakage past piston rings or valve guides (4-stroke engine), or excess lubricant (2-stroke engine)

Overheating - A blistered white insulator and glazed electrodes. Caused by ignition system fault, incorrect fuel, or cooling system fault

Worn plug - Worn electrodes will cause poor starting in damp or cold weather and will also waste fuel

10 Fault diagnosis: ignition system

Symptom	Cause	Remedy
Engine will not start	Faulty ignition switch	Operate switch several times in case contacts are dirty. If lights and other electrics function, switch may need renewal
	Faulty CDI unit	Have unit tested. Replace if required
	Wiring fault	Check and repair wiring
	Faulty pulser generator	Test pulser coils. Renew if required
	Starter motor not working	Discharged battery. Remove battery from machine and recharge
		Faulty starter circuit. Check for continuity
	Short circuit in wiring	Check whether fuse is intact. Eliminate fault before switching on again
	Completely discharged battery	If lights do not work, remove battery and recharge
Engine misfires	Faulty ignition coil	Test and renew if required
	Wiring fault	Check and repair wiring
	Faulty HT lead	Renew lead
	Faulty sparking plug	Renew plug or clean if fouled
	Faulty pulser coil	Test pulser coils. Renew pulser generator if required
Engine lacks power and overheats	Retarded ignition timing due to ATU unit failure	Check ATU
	Vacuum advance faulty	Test and renew if required
Engine 'fades' when under load	Pre-ignition	Check grade of plugs fitted; use recommended grades only

Chapter 5 Frame and forks

Contents

Specifications

Frame type .... Double cradle

Front forks

Make	Showa
Type	Air-assisted, hydraulically damped, telescopic
Travel	147 mm (5.8 in)
Air pressure	14 - 21 psi (1.0 - 1.5 kg/cm^2)
Oil capacity:	
USA	240 cc (8.11 US fl oz)
UK	220 cc (7.74 Imp fl oz)
Spring free length:	
Upper	111.7 mm (4.40 in)
Service limit	106.0 mm (4.17 in)
Lower	466.2 mm (18.35 in)
Service limit	442.0 mm (17.40 in)
Stanchion OD	38.950 - 38.975 mm (1.5335 - 1.5344 in)
Service limit	38.9 mm (1.5315 in)
Bush I D	39.02 - 39.11 mm (1.536 - 1.540 in)
Service limit	39.16 mm (1.542 in)
Lower leg I D	40.040 - 40.080 mm (1.5764 - 1.5779 in)
Service limit	40.130 mm (1.5799 in)

Rear suspension .... Swinging arm, controlled by two air assisted, hydraulically damped suspension units

Rear suspension units

Make	Showa
Type	Air assisted, hydraulically damped
Travel	80 mm (3.2 in)
Air pressure:	
UK and 1980 (US) models	28 - 42 psi (2.0 - 3.0 kg/cm^2)
1981 (US) models	28 - 57 psi (2.0 - 4.0 kg/cm^2)
Oil capacity	365 cc (12.34/12.85 US/Imp fl oz)

Torque wrench settings

	kgf m	lbf ft
Steering stem nut (24 mm)	8.0 - 12.0	58 - 87
Handlebar clamp bolts (8 mm)	1.8 - 2.5	13 - 18
Top yoke pinch bolt nut (8 mm)	1.8 - 2.5	13 - 18
Top yoke pinch bolt nuts (7 mm)	0.9 - 1.3	7 - 10
Lower yoke pinch bolts (8 mm)	3.0 - 4.0	22 - 29
Fork top bolts (34 mm)	1.5 - 3.0	11 - 22
Air valve (8 mm)	0.4 - 0.7	3 - 5
Air hose connector (8 mm)	0.4 - 0.7	3 - 5
Fork joint hose (10 mm)	1.5 - 2.0	11 - 14
Fork joint hose (8 mm)	0.4 - 0.7	3 - 5
Rear suspension unit securing bolt nuts (10 mm)	3.0 - 4.0	22 - 29
Air hose connector (to 3-way joint) (8 mm)	0.8 - 1.2	6 - 9
Air hose joint (10 mm)	1.5 - 2.0	11 - 14
Air hose joint (8 mm)	0.4 - 0.7	3 - 5
Air pressure sensor	0.8 - 1.2	6 - 9
Swinging arm right-hand pivot bolt (30 mm)	8.0 - 12.0	58 - 87
Swinging arm left-hand pivot bolt (30 mm)	1.8	13
Swinging arm pivot bolt locknut (30 mm)	8.0 - 12.0	58 - 87
Final drive bevel housing securing nuts (10 mm)	3.5 - 4.5	25 - 33
Headlamp casing mounting bolts (10 mm)	1.8 - 2.5	13 - 18
Main stand pivot bolt nuts (10 mm)	2.0 - 2.4	11 - 17
Prop stand pivot bolt nut (10 mm)	2.0 - 2.4	11 - 17

Interstate models only

Horn retaining bolts (8 mm)	1.2	9
Frame to mounting bracket bolt (8 mm)	2.5	18
Frame to mounting bracket nuts (6 mm)	1.2	9

1 General description

The frame utilized on the Honda GL 1100 is of the full-cradle type; that is, the engine does not comprise any part of the frame. The massive frame incorporates duplex down tubes which run from the steering head lug and either side of the engine, to a point to the rear of the engine, and are additionally strengthened by a tubular cross-member forward of the engine. The frame top tubes and cross tubes are also duplicated, the latter being extended to form a rear to which the dualseat and rear mudguard are fixed. The frame down tube on the left of the machine is fitted with a detachable sub-frame in the horizontal section, which can be removed to facilitate engine removal.

The suspension, however, does not follow convention to the same extent as the frame. The front forks, whilst being of the usual telescopic type, have unusual internal damping arrangements. Although the damping medium is still predominently oil, air pressure is used as a variable pre-loading force. Conventional fork springs are contained within the fork stanchions. Air is added to the forks by means of a Shrader valve, enclosed beneath a chromed protector cap, at the top of the right-hand fork leg.

The rear suspension units operate on the same principle as the front forks and can be dismantled to the point where the rubber gaiters, oil seals and bushes are separated from the main unit casings. Air is added to both units by means of a Shrader valve, located beneath a chromed protecter cap, just above the right-hand pillion footrest. Should the air pressure in these units fall below a set minimum pressure, a warning light in the face of the tachometer will illuminate. Honda recommended that once this happens, the speed of the machine should be reduced to below 50 mph and air added to the units at the earliest possible opportunity.

Because air is used as a variable pre-loading force, the damping characteristics of both the front forks and rear suspension units can be adjusted almost infinitely over the prescribed range to suit both riding conditions and machine loading. To avoid differing pressures in each fork leg, the two legs are interconnected; this applies also to the rear suspension units.

Rear suspension is of the swinging arm type, using the above mentioned units to provide the necessary damping action. The swinging arm is constructed of square-section tubing, the right-hand arm member serving also as a torque tube through which the final drive shaft passes and to which the final drive gear casing is attached. The fork assembly pivots on needle roller bearings fitted either side of the tubular cross member.

2 Front forks: methods of removal

1   If necessary, it is possible to remove the forks as an assembly, together with the lower yoke. It is unlikely that this method will prove advantageous in view of the amount of preliminary dismantling necessary, but if this approach is considered to be necessary, due to there being frontal damage or the steering head assembly requiring attention, follow the sequence detailed in Section 5, leaving the stanchions clamped in the lower yoke.

2   In most cases, the forks are best removed individually, without disturbing the yokes or steering head bearings. The relevant sequence is described in Section 3 of this Chapter.

3 Front fork legs: removal and refitting

1   Place the machine securely on its centre stand, leaving adequate room around the front wheel area for comfortable working. It should be noted at this stage that some means of supporting the front wheel clear of the ground must be arranged. This can be accomplished by placing blocks or a jack beneath the sump, taking care not to damage the delicate sump fins. Alternatively a pair of tie-down straps, of the type used to secure motorcycles on trailers, can be arranged to lash the rear of the machine to the ground. Whatever method is chosen, make sure that there is no risk of the machine falling over whilst the front wheel and forks are removed.

2   Refer to Section 3 of this Chapter 6 and remove the front wheel. Remove the front mudguard by unscrewing the four bolts which secure it to the fork lower legs. Lower the mudguard away from the fork legs, threading the speedometer cable through the guide and unclipping the brake hoses from

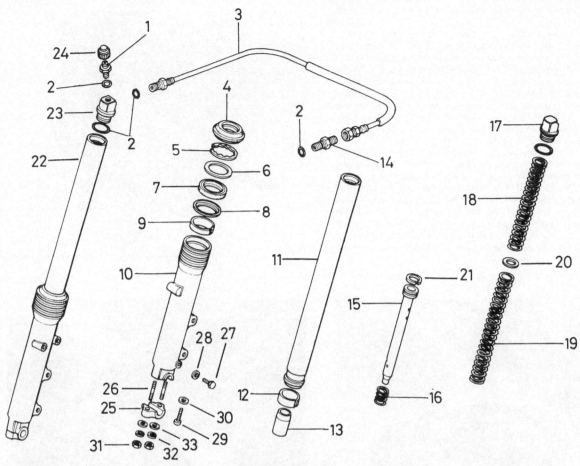

Fig. 5.1 Front forks

1 Air valve	10 Lower leg	18 Upper spring	26 Stud – 2 off
2 O-ring	11 Stanchion	19 Lower spring	27 Drain plug
3 Air hose	12 Bush	20 Separator plate	28 Drain plug gasket
4 Dust seal	13 Damper rod seat	21 Damper rod piston ring	29 Allen bolt
5 Circlip	14 Union – 2 off	22 Right-hand fork leg	30 Washer
6 Backing ring	15 Damper rod	23 Fork top bolt	31 Nut – 2 off
7 Oil seal	16 Rebound spring	24 Protector cap	32 Spring washer – 2 off
8 Backing ring	17 Fork top bolt	25 Wheel spindle clamp	33 Washer – 2 off
9 Bush			

their retaining clips before doing so.

3   Remove the chromed protector cap from the Shrader valve on the cap of the right-hand fork leg and release the air pressure from both fork legs by depressing the centre of the valve. Disconnect the air hose from first the left-hand leg and then the right-end leg; disconnecting in this order will obviate any chance of the hose twisting along its length.

4   Remove each fork leg by loosening the upper and lower yoke clamp bolts and easing the leg downwards, out of position. Note that if it is intended to dismantle the fork legs, the fork cap should be slackened prior to loosening the yoke bolts. This may be done by fitting an open-ended spanner over the flats of the cap, ensuring that the air hose is moved clear of the cap connection before turning the cap.

5   The fork legs can be refitted by reversing the sequence given for removal. If the legs have been drained of oil whilst removed, they should be located in the fork yokes and refilled with the specified quantity of automatic transmission fluid (ATF) by one of the following methods.

6   Remove the hose adaptor from the left-hand fork cap and turn each fork stanchion so the threaded hole in each cap faces outwards; the specified amount of ATF may then be poured into each fork leg through the cap holes. Refit the hose adaptor,

renewing and greasing the O-ring before doing so. Tighten the adaptor to a torque of 0.4 - 0.7 kgf m (3.5 lbf ft).

7   The alternative method of refilling the fork legs requires the use of a syringe. With the correct quantity of ATF in the syringe, remove the drain plug and sealing washer from each lower leg and inject the fluid into each fork leg through the drain hole. Remove the syringe and quickly refit the drain plug after having first checked the condition of its sealing washer and renewed it if necessary.

8   Ensure that before the clamp bolts are tightened, the centre of the threaded hole in each fork cap aligns with the punch mark on the face of the upper yoke. Torque load the upper yoke clamp bolt nuts to 0.9 - 1.3 kgf m (7 - 10 lbf ft) and the lower yoke clamp bolts to 3.0 - 4.0 kgf m (22 - 29 lbf ft).

9   Reconnect the air hose to the right-hand fork leg cap, renewing and greasing the O-ring before doing so. Tighten the hose connections to a torque of 0.4 - 0.7 kgf m (3 - 5 lbf ft). Reconnect the other end of the hose to the adaptor of the left-hand fork leg cap, tightening it to a torque of 1.5 - 2.0 kgf m (11 - 14 lbf ft). A special 'crowsfoot' adaptor will be needed to enable a torque wrench to be used on the hose ends. If this tool is not available, the connections should be screwed in until finger-tight and then nipped tight with an open-ended spanner.

Do not overtighten the connections; there is a high risk of their shearing.

10 With the front mudgaurd and wheel refitted, refill the fork legs with air to a pressure of 14 - 21 psi (1.0 - 1.5 kg/sq cm). Do not use a compressed air supply to do this because there is a risk of dislodging and damaging the fork leg seals. One of the various types of mini syringe pumps now available at motor-cycle dealers is ideal for this purpose. These pumps are as little as five inches long and can therefore easily be carried on the machine; they come supplied with flexible connectors. The maximum pressure the fork legs can withstand before damage occurs is 43 psi (3 kg/sq cm).

11 Once charged, take the machine off its stand and with the front brake applied, pump the forks up and down several times. Place the machine back on its centre stand and check the air pressure on the forks. If this pressure is within the specified limits and that required by the rider, the chromed protector cap may be refitted to the Shrader valve. It should be noted that each time the pressure gauge is removed from the valve, a small volume of air will be lost from the fork legs resulting in an equally small drop in pressure. This should be allowed for when noting the gauge reading.

3.2 The front mudguard is retained to each fork leg by two bolts

3.4a Loosen the upper yoke clamp bolts ...

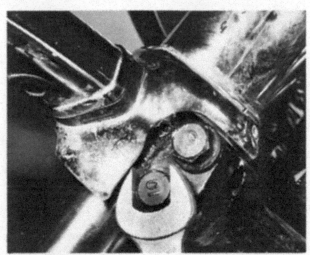

3.4b ... loosen the lower yoke clamp bolts ...

3.4c ... and ease the fork leg out of position

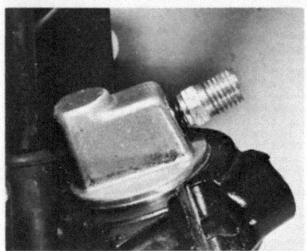

3.6 Refit the air hose adaptor and O-ring

3.7 Fluid may be injected into each fork leg through the drain hole

3.9 Ensure the air hose connections are correctly positioned before tightening

4 Front fork legs: dismantling, examination and re-assembly

1   Having removed the fork legs as described in Section 3, they may be dismantled for further examination. Always deal with one leg at a time and on no account interchange components from one leg to the other as the various moving parts will have bedded in during use, and should remain matched. Commence by draining the oil, either by way of the drain plug in the lower leg or by removing the top cap and inverting the leg. Pumping the unit will assist in the draining operation. The cap is spring loaded quite heavily by the fork springs, so care should be taken when unscrewing it.
2   Refer to the accompanying line drawing then start dismantling, laying each part out on a clean surface as it is removed. With the cap removed, the two fork springs and their separator plate will slide out of the fork stanchion as it is inverted. Place the stanchion between the padded jaws of a vice and tighten the vice just enough to hold securely the fork leg in place. Unscrew and remove the Allen key bolt and washer from the underside of the fork lower leg. Detach the dust seal and remove the circlip located beneath it. The stanchion and lower leg can now be separated by using the lower leg as a slide hammer by which to dislodge the bush, together with the backing rings and oil seal, from its seating. In practice, some force and a considerable number of attempts were needed to achieve this. It must be emphasised, however, that care must be taken not to overtighten the vice jaws or cause damage to the alloy of the lower leg. Once separated, the stanchion, lower leg and damper assembly, together with the bushes, seals and backing rings, may be laid out on the work surface and examined as follows.
3   It is not generally possible to straighten forks which have been badly damaged in an accident, particularly when the correct jigs are not available. It is always best to err on the side of safety and fit new ones, especially since there is no easy means to detect whether the forks have been overstressed or metal fatigued. Fork stanchion (tubes) can be checked, after removal from the lower legs, by rolling them on a dead flat surface. Any misalignment will be immediately obvious.
4   The fork springs will take a permanent set after considerable usage and will need renewal if the fork action becomes spongy. The service limit for the total free length of each spring is given in the Specifications. Always renew them as matched pairs.
5   The parts most liable to wear over an extended period of

time are the internal surfaces of the lower leg, the lower leg bush and the outer surfaces of the fork stanchion. The limits of wear allowed on these components are listed in the Specifications of this Chapter. Refit the bush into the lower leg before measuring its internal diameter.
6   Inspect the fork stanchion, lower leg, damper assembly and bushes for scoring and renew each component as necessary. Scoring over the length of the stanchion where it enters the oil seal will damage the seal and lead to fluid leakage.
7   It is advisable to renew the oil seals when the forks are dismantled even if they appear to be in good condition. This will save a strip-down of the forks at a later date if oil leakage occurs.
8   Check that the dust excluder rubbers are not split or worn where they bear on the fork tube. A worn excluder will allow the ingress of dust and water which will damage the oil seal and eventually cause wear of the fork tube.
9   If damping action is lost, the piston around the damper piston should be renewed; it is catalogued as a separate item. Check that the small holes in the damper tubes are not blocked and if no substantial improvement is shown when the forks are reassembled and refilled, renew the complete damper assembly.
10 Reassembly of the fork legs is essentially a reversal of the sequence given for dismantling, noting the following points. Ensure that all component parts have been thoroughly cleaned before commencing reassembly. The small rebound spring and damper assembly should be inserted into the fork stanchion, followed by the long fork spring, separator plate and short fork spring. Note that the close coil end of the long fork spring must be facing towards the top of the stanchion. Inspect the O-ring fitted to the fork cap for damage or deterioration and renew if necessary. Refit the cap to the stanchion by gripping the flats of the cap between the protected jaws of the vice, locating the end of the fork spring against the base of the cap and both pushing and turning the stanchion against the cap so as to compress the spring and locate the threads of the two components. Reposition the assembly in the vice jaws so that the stanchion is gripped; tighten the cap to the recommended torque of 1.5 - 3.0 kgf m (11 - 22 lbf ft).
11 The damper rod seat and fork stanchion assembly should now be inserted into the lower leg and secured in position with the Allen bolt. Before fitting the Allen bolt, check the condition of its sealing washer and renew it if thought necessary. Clean the thread of the bolt and coat it with locking fluid. Refit the bolt and tighten it to the recommended torque of 1.5 -2.5 kgf m (11 - 18 lbf ft).

12 With the fork assembly positioned upright on the work-surface, or supported between the protected jaws of the vice, fit the bush into the fork lower leg by passing it down over the fork stanchion and carefully drifting it into position with a length of tube of an internal diameter slightly larger than that of the stanchion and an external diameter slightly less than that of the bush. Ensure that the end of this tube is both square to its sides and free of burrs.

13 Refit the backing ring over the bush followed by the new oil seal. Before fitting the seal, lubricate it thoroughly in ATF and ensure it is correctly positioned over the stanchion with the marked face uppermost. The seal can be drifted into position using the method described for fitting the bush, although Honda supply tool no 07947 - 4630100 specifically for this purpose. The seal is in its correct fitted position once the circlip groove appears above its upper surface.

14 Finally, refit the backing ring, circlip and dust seal; noting that the rounded face of the circlip should face downwards. The fork legs should be refilled with the correct quantity of ATF once they are refitted to the machine.

4.6 Inspect the fork bushes for scoring

4.9 Check that the small holes in the damper tube are not blocked

4.10a Insert the damper assembly into the fork stanchion ...

4.10b ... followed by the long fork spring ...

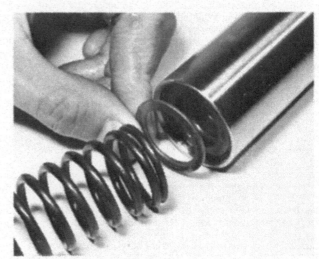

4.10c ... the separator plate and the shorter fork spring

4.10d Inspect the condition of the fork cap O-ring before fitting the cap to the stanchion

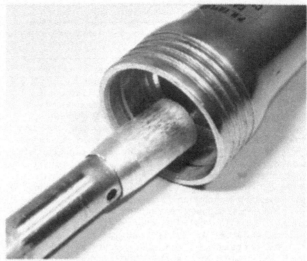

4.11a Fit the damper rod seat and insert the stanchion assembly into the fork lower leg

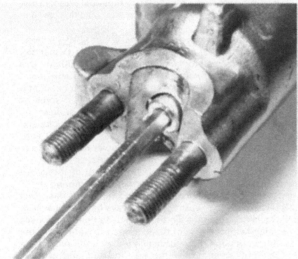

4.11b Secure the stanchion to the lower leg with the Allen bolt

4.12 Fit the fork bush and seals in the order shown

4.14a Ensure that the circlip is correctly located ...

4.14b ... before fitting the dust seal

5 Fork yokes and steering head bearings: removal and refitting

1   Fork yoke dismantling is an unwieldy process which is best avoided if at all possible. A considerable amount of work is involved in removing the numerous appendages from the fork yokes. For this reason, fork removal is best carried out as described in Section 3. It should be noted that some operations, such as inspection of the steering head bearings, can be undertaken without resorting to the full strip-down sequence given here, the trick being to disturb only those components which directly affect the part-dismantling sequence required. Note that on Interstate models the fairing should be detached to give improved access (see Section 19) and the following procedure modified as common sense dictates. Before disconnecting any electrical connections the battery must be isolated by removing the lead from the positive terminal; this will prevent shorting out and the resultant damage to components.

2   Start by removing the front wheel and fork legs as described in Section 3 of Chapter 6 and Section 3 of this Chapter respectively. Although it is possible to leave the fork legs attached to the lower yoke, this makes an awkward job all the more difficult; they are best removed. The handlebar assembly must be removed, either completely after releasing the various switch harnesses, hydraulic pipe and cables, or by removing the handlebar clamp and displacing the assembly to provide clearance. In each case it is best to either to remove the dummy fuel tank or to protect its surface against damage by covering it with a pad of clean rags or a protective mat; the displaced handlebar assembly may then be rested on the tank. Note that the hydraulic front brake master cylinder/reservoir can be detached from the handlebar, complete with control lever, without disconnecting the hydraulic hose. Remove the control clamp, which is retained by two bolts, and lift the complete brake operating assembly away from the bars. Take care not to tip the reservoir so that the hydraulic fluid is not spilt. Brake fluid is a very effective paint stripper, and will damage plastic components. Tie the brake cylinder assembly to some part of the machine that is not to be disturbed.

3   Remove the head lamp reflector unit, complete with rim, so that access may be made to the main wiring connections. The rim is retained by two cross-head screws passing through the shell at the 4 o'clock and 8 o'clock positions. Disconnect all the wires that lead to the handlebar switches, the instruments and the main harness. Pull the freed wires through the rear of the shell. No difficulty should be encountered when reconnecting wires as they are colour-coded. If there is any doubt, mark the particular wires with tape so that they may be identified later.

4   Detach the headlamp shell from the mounting 'ears', which extend from the fork shrouds. Unscrew the reflectors and pull off the reflector seats to enable a spanner to be applied to the headlamp retaining bolts.

5   Disconnect the speedometer and tachometer drive cables at the instrument heads by unscrewing the knurled ring on each cable end. Support the instruments and remove the two nuts which retain the instrument mounting bracket to the fork upper yoke. The complete assembly, including the warning lamp console, can be lifted from position.

6   Disconnect the main leads to the ignition switch at the block connector. Remove the ignition switch, after unscrewing the two bolts which pass through the switch flange from the underside.

7   Prise the rubber cap from the top of the steering stem. Using a C-spanner, remove the slotted nut, followed by the plain washer, from the steering stem. Loosen the pinch bolt which passes horizontally through the rear of the upper (crown) yoke. With the aid of a soft-faced mallet, the upper yoke can be tapped upwards and off the steering stem. At the same time as the yoke is moved upwards, the two fork shrouds, complete with the flashing indicators and shroud guides, can be removed. Knock back the tabs of the lockwasher from the slotted locknut and using a C-spanner, loosen and remove the locknut followed by the lockwasher.

8   Unbolt the hydraulic hose two-way joint from the lower yoke. Detach the previously tied brake components and lift the interconnected components away from the machine as one assembly.

9   To release the lower yoke and the steering head stem, unscrew the adjuster ring at the top with a suitable C-spanner. Honda supply a special tool no 07916 - 3710100 for this purpose. Alternatively, if neither tool is available, a soft metal drift and hammer may be used to loosen the ring. It should be noted, however, that during assembly a pegged sprocket spanner will be required to torque accurately the bearing to the prescribed pre-load. As the nut is removed, the lower yoke and steering stem will drop downwards and thus should be supported until the nut is removed. Carefully lower and disengage the lower yoke assembly, placing it to one side.

10 Reassembly of the components is a reversal of the above sequence, noting the following points. The steering head bearings should be checked, greased and adjusted as described in Section 6. A new lockwasher must always be used, never fit the used washer. Tighten the slotted locknut so that the grooves in the nut align with the tabs of the lockwasher. Note that if the grooves in the nut cannot be easily aligned with the tabs of the washer, the nut should be removed, inverted and refitted. Torque load the upper slotted nut to 8.0 - 12.0 kgf m (58 - 87 lbf ft) before refitting the rubber cap to the top of the steering stem. Tighten the pinch bolt to a torque of 1.8 - 2.5 kgf m (13 - 18 lbf ft).

5.3 Disconnect all electrical connections within the headlamp shell

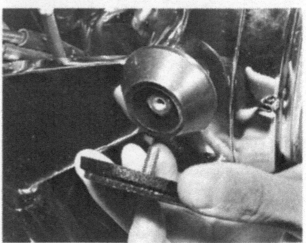

5.4 The headlamp retaining bolts are located beneath the reflector seats

5.5 Disconnect each instrument cable by unscrewing the retaining ring

5.8 Unbolt the hydraulic hose 2-way joint from the lower yoke

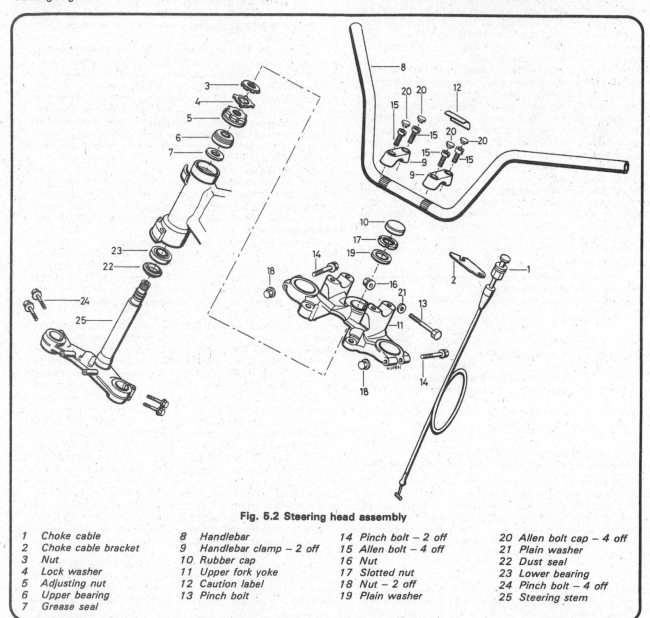

Fig. 5.2 Steering head assembly

1	Choke cable	8	Handlebar	14	Pinch bolt – 2 off
2	Choke cable bracket	9	Handlebar clamp – 2 off	15	Allen bolt – 4 off
3	Nut	10	Rubber cap	16	Nut
4	Lock washer	11	Upper fork yoke	17	Slotted nut
5	Adjusting nut	12	Caution label	18	Nut – 2 off
6	Upper bearing	13	Pinch bolt	19	Plain washer
7	Grease seal				

20	Allen bolt cap – 4 off
21	Plain washer
22	Dust seal
23	Lower bearing
24	Pinch bolt – 4 off
25	Steering stem

6   Steering head bearings: examination and adjustment

1   The steering head bearings are of the taper roller variety, and are unlikely to give rise to problems in the normal life of the machine, although the manufacturer recommends examination, lubrication and adjustment of these components at 8000 mile intervals. Access to the bearings is gained by following the procedure detailed in Section 5 of this Chapter.

2   It will be noted that the bearings are effectively in two parts; the outer races, which will remain in the steering head tube, and the inner race, cage and rollers, which will come away as the lower yoke is removed. Of the latter, the lower bearing will probably be firmly attached to the steering stem, whilst the upper bearing will lift away quite easily. It is normally possible to lever the lower bearing off the steering stem, but it may prove necessary to employ a bearing extractor in stubborn cases.

3   The outer races can be removed with the aid of a long drift, passed through the steering head tube, new races being fitted by judicious use of a tubular drift, such as a large socket or similar. Before any decision is made to remove the outer races, they should be cleaned and checked as described below.

4   Wash out the bearings with clean petrol to remove all traces of old grease or dirt. Check the faces of the rollers and the outer races for signs of wear or pitting, both of which are unlikely unless the machine has been neglected in the past. If damaged, the bearings must be renewed.

5   When assembling the steering head, pack each bearing with grease prior to installation. The lower yoke is offered up and the bearing and adjuster nut is fitted finger tight. Adjustment of the bearings requires a socket-type peg spanner so that the slotted adjuster nut can be set to the prescribed torque loading. This requires the use of the special Honda steering stem socket (Part No 07916 - 3710100) or a home-made equivalent. A piece of tubing can be filed to fit the nut and then welded to a damaged socket to improvise.

6   Tighten the adjuster nut to a torque of 1.4 - 1.6 kgf m (10 - 12 lbf ft) and turn the steering stem from lock to lock five times to allow the bearings to seat properly. Repeat this sequence twice. Should the nut fail to tighten after turning the steering stem the first or second time, remove the nut and inspect both the stem and nut threads for burrs or contamination by dirt or grit. Obviously, rough or dirty threads will prevent correct torque loading of the bearing itself.

7   Steering lock: maintenance

1   A security lock is mounted on the headstock, enabling the owner to immobilise the machine by locking the steering in one position. The lock consists of a key operated plunger which engages in a slot in the steering head. A small return spring disengages the lock mechanism when the key is released.

2   Maintenance is confined to keeping the lock lightly lubricated using light machine oil or one of the multipurpose aerosol lubricants. In the event that the lock malfunctions, it will be necessary to remove the body, after removal of the retaining rivet and washer, and to fit a replacement unit.

8   Frame: examination and renovation

1   The frame is unlikely to require attention unless accident damage had occurred. In some cases, renewal of the frame is the only satisfactory remedy if the frame is badly out of alignment. Only a few frame specialists have the jigs and mandrels necessary for resetting the frame to the required standard of accuracy, and even then there is no easy means of assessing to what extent the frame may have been overstressed.

2   If the front forks have been removed from the machine, it is comparatively simple to make a quick visual check of alignment by inserting a long tube that is a good push fit in the steering head races. Viewed from the front, the tube should line up exactly with the centre line of the frame. Any marked deviation from the true vertical position will immediately be obvious; the steering head is a particularly good guide to the correctness of alignment when front end damage has occurred.

3   Remember that a frame which is out of alignment will cause handling problems and may even promote 'speed wobbles'. If misalignment is suspected, as a result of an accident, it will be necessary to strip the machine completely so that the frame can be checked, and if necessary, renewed.

4   After the machine has covered a considerable mileage, it is advisable to examine the frame closely for signs of cracking or splitting at the welded joints. Rust corrosion can also cause weakness at these joints. Minor damage can be repaired by welding or brazing, depending on the extent and nature of the damage.

9   Swinging arm: removal, examination, renovation and refitting

1   The swinging arm fork pivots on tapered roller bearings fitted one either side of the fork tubular cross member. These bearings are mounted on adjustable screw pivot stubs fitted into the frame lugs located forward of the frame cradle tubes and downtubes.

2   Any wear in the swinging arm bearings is characterised by a tendency for the rear of the machine to twitch when ridden hard through a series of bends. Wear can be checked by placing the machine on the centre stand, with the rear wheel clear of the ground, and pushing the swinging arm fork ends from side to side. Any discernible free play will indicate the need for adjustment as described in Routine Maintenance. If attempts at adjustment prove unsuccessful due to excessive wear or bearing damage removal of the swinging arm for further examination will be required.

3   Remove the rear wheel by following the procedure given in Section 11 of Chapter 6. Temporarily refit the left-hand suspension unit lower bolt so that the weight of the gear housing is taken. Loosen and remove the three nuts which retain the final drive housing to the torque tube flange. Lift the housing away.

4   Remove the external circlip from the inside of the splined joint at the output end of the final drive shaft. The circlip is deeply recessed so will require a pair of long-jawed circlip pliers for easy removal. Pull the splined joint off the shaft. Prise the rubber boot from the engine casing lip and remove the external circlip that retains the U-joint to the splined output shaft. Pull the joint off the splines.

5   Free the rear brake hydraulic hose from the clips on the left-hand fork arm and prise the plastic caps off the swinging arm pivot stubs ends. The left-hand stub is locked by means of a slotted lockring. Loosening the ring may be accomplished, at this stage, using a soft-metal drift. It should be noted, however, that a fabricated tool, as described in paragraph 10, must be used to tighten the ring on reassembly. If the tool is fabricated now it may be used to loosen the ring in preference to using a drift. Using a suitable Allen key on the left-hand stub and a spanner on the right, slowly unscrew both pivot stubs, taking the weight of the swinging arm assembly as the pivot stubs leave the bearings. Lift the swinging arm fork out towards the rear of the machine and with it placed on the work surface, pull the final drive shaft out of the torque tube from the forward end. Check the condition of the dust seal remaining in the rearmost end of the torque tube; if it is damaged in any way it should be renewed. The seal may be removed by carefully levering it out of position with the flat of a screwdriver. Care should be taken not to damage the mating surfaces of the torque tube during removal of the seal.

6   Remove the dust seals from each side of the fork crossmember by levering them out with the flat of a screwdriver. The taper roller bearings may now be lifted out of their location in each end of the cross-member and washed in petrol so that all

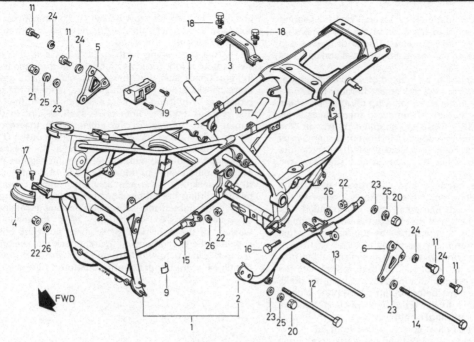

Fig. 5.3 Frame

1	Frame assembly	10	Caution label	19	Screw – 2 off	
2	Detachable side member	11	Bolt – 4 off	20	Domed nut	
3	Seat brace	12	Bolt	21	Domed nut	
4	Cover	13	Bolt	22	Nut – 2 off A1 model (3 off A model)	
5	Right-hand engine plate	14	Bolt			
6	Left-hand engine plate	15	Bolt	23	Washer – 5 off	
7	Helmet holder	16	Bolt	24	Spring washer – 4 off	
8	Caution label	17	Bolt – 2 off	25	Spring washer – 4 off	
9	Caution label	18	Bolt – 2 off	26	Spring washer – 2 off A1 model (3 off A model)	

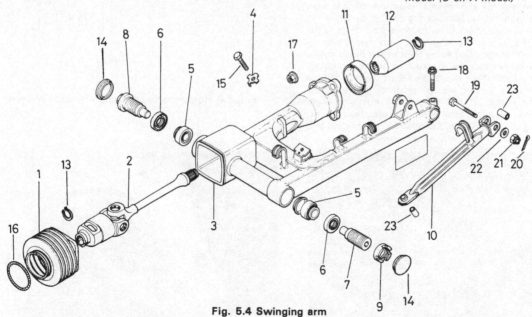

Fig. 5.4 Swinging arm

1	Shaft boot	9	Adjuster lock ring	17	Nut – 3 off
2	Final drive shaft	10	Brake torque arm	18	Bolt
3	Swinging arm	11	Dust seal	19	Bolt
4	Tab washer	12	Coupling sleeve	20	Split pin
5	Taper roller bearing – 2 off	13	Circlip	21	Castellated nut
6	Dust seal – 2 off	14	End cap – 2 off	22	Washer
7	Left-hand pivot stud	15	Bolt	23	Spacer
8	Right-hand pivot stud	16	Boot retainer		

old grease and any contamination is removed. Wipe the bearing faces of the outer races left in the ends of the cross member with a petrol dampened rag so that all traces of grease and contamination are removed and inspect both races and bearings for signs of damage or wear.

7   If the bearings are found to be in need of renewal, the outer races must be extracted from the fork cross-member by using a slide hammer as shown in the accompanying photograph. There is no other acceptable method of removing these races apart from the tool no 07936 - 8890100 supplied by Honda. The new races may be drifted into position in the cross-member by using a length of metal tube or a socket of the appropriate diameter whilst ensuring that the race is kept square to the cross-member at all times.

8   Before fitting the bearings into the races, pack them liberally with high melting point grease. Apply grease to the inside face of the dust seals and press them into position over the bearings. It should be noted that this is the only method by which the swinging arm bearings may be lubricated; it must therefore be done as thoroughly as possible.

9   Refit the swinging arm by reversing the dismantling procedure and by following the directions that follow. Lift the swinging arm fork into position in the frame, routing the brake hose between the fork arms before doing so. Apply a blob of high melting point grease to the tip of each pivot stub before fitting and tightening them finger-tight.

10   Tighten the right-hand pivot stub to a torque of 8.0 - 12.0 kgf m (58 -87 lbf ft). Tighten the left-hand pivot stub to a torque of 1.8 kgf m (13 lbf ft), using an adaptor made up of a length of 10 mm Allen key fitted into a socket or the Honda tool no 07917 - 3710000. Move the swinging arm fork through its operating arc several times and retighten the left-hand pivot stub to the correct torque loading. Retain the left-hand pivot stub in position by fitting the lockring and tightening it to a torque of 8.0 - 12.0 kgf m (58 - 87 lbf ft), holding the pivot stub in position to prevent it from turning. If the special Honda tool no 07908 - 4690001 is not available for this purpose, a tool with which to tighten the lockring may be manufactured by using a short length of thick-walled tubing and a nut with the same overall diameter as the internal diameter of the tube. Refer to the accompanying illustration for details, cutting away the segments shown with a hacksaw to leave four tangs. If the machine is to be regarded as a long term purchase, it may be considered worthwhile spending some time with a file to obtain a good fit. The end can then be heated to a cherry red colour and quenched in oil to harden the tangs. Insert the nut into the other end of the tube, leaving enough protruding to ensure a good fit into the socket. It will help if the nut is a slight interference fit on the tube otherwise difficulty may be experienced holding it in the required position. Weld the nut into

position. The tool may now be used in conjunction with a socket and torque wrench to tighten the nut, checking frequently that the pivot stub is not moving.

11   Before proceeding further, carry out a final check on the swinging arm fork for smooth operation throughout its operating arc. Check also that there is no side to side movement. If these checks prove to be satisfactory, carry on the reassembly procedure by refitting the final drive housing to the torque tube and reconnecting the final drive shaft U-joint to the splined output shaft, lubricating the splines with the recommended grease before doing so . Note the condition of the O-ring of the housing bearing retainer and renew it if necessary, before engaging the housing with the torque tube. Tighten the housing securing nuts evenly to a torque of 3.5 - 4.5 kgf m (25 - 33 lbf ft) and check that the U-joint securing circlip is correctly located before refitting the rubber boot to the engine casing lip.

12   Fit and tighten the suspension unit lower retaining nuts to a torque of 3.0 - 4.0 kgf m (22 - 29 lbf ft), ensuring that the washers of the right-hand attachment point are positioned as shown in the accompanying figure. Check that the brake hose is securely clipped to the left-hand fork arm and the pivot stub plastic caps are correctly refitted before fitting the rear wheel.

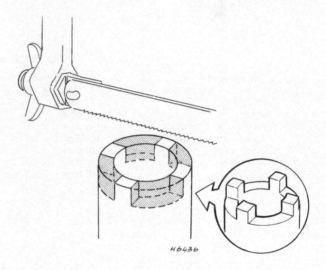

H6436

Fig. 5.5 Home-made swinging arm stub spanner

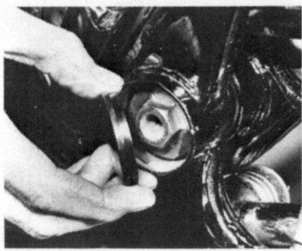

9.5a Remove the plastic caps from the swinging arm pivot ends

9.5b Prise the dust seal out of the torque tube end

9.7a Use a slide hammer to remove the bearing races

9.7b Insert new races into the cross member ...

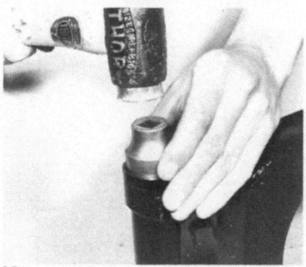

9.7c ... and carefully drift them into position

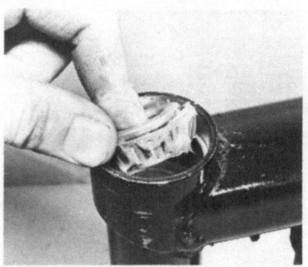

9.8a Pack the bearings with high melting point grease

9.8b Press the dust seal into position over the bearing

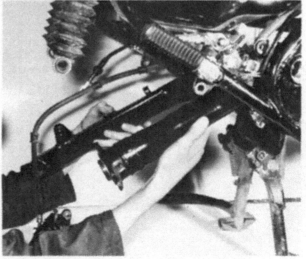

9.9a Lift the swinging arm into position in the frame ...

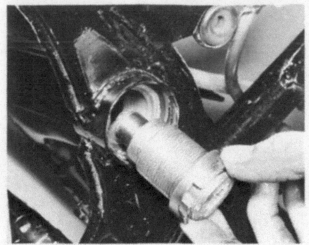

9.9b ... and refit the pivot stubs

9.10a Torque load the right-hand pivot stub ...

9.10b ... followed by the left-hand pivot stub and locknut

9.11a Note the condition of the bearing retainer O-ring ...

10 Rear suspension units: removal, examination, reno-vation and refitting

1　Before removal of either suspension unit may take place, the seat and right-hand sidepanel must be removed to give access to the air charging point/pressure switch assembly and the air hose connections. Remove the chromed protector cap from the Shrader valve and release the air pressure from both suspension units by depressing the centre of the valve.

2　Pull apart the air pressure sensor electrical connection after having first unclipped it from the air charging connection mounting plate. Remove the large nut and plain backing washer from the charging point and ease the 3-way union clear of the mounting plate and clear of the machine. Disconnect the two air hoses from the union connections and store the union in a clean place until it is required for refitting.

3　Remove the grab rail from its frame and rear mudguard attachments by unscrewing and removing the suspension unit upper attachment securing nuts and washers and the three rail to mudguard securing bolts and washers. Spring the ends of the rail apart so that they clear the suspension unit mounting stubs and lift the rail clear of the machine. Unscrew and remove the left-hand unit lower mounting bolt and the right-hand unit lower mounting nut and washer; support the weight of the rear wheel and pull the units clear of the machine, taking care to route the

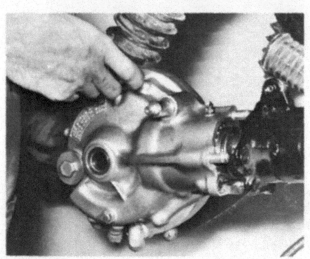

9.11b ... before fitting the housing to the torque tube

air hoses clear of the frame at the same time. Service each unit as follows.

4   Remove the rubber gaiter from the unit and inspect it for signs of perishing or splitting. The gaiter should be renewed if damaged in any way as it prevents road dirt and moisture from finding its way to and destroying the oil seal. Remove the air hose from the unit and inspect the condition of both the hose rubber and the O-ring fitted over the union. If either component is damaged or perished it should be renewed although Honda recommended that the O-ring be renewed whenever the hose is disconnected.

5   Using a pair of circlip pliers, remove the circlip from the unit casing. The backing ring can now be hooked out of the casing with the flat of a small screwdriver to expose the oil seal located beneath it.

6   Removal of the oil seal requires some degree of care and the use of a compressed air supply. It should be remembered that the suspension unit contains 365 cc of automatic transmission fluid (ATF) and once the seal is dislodged, this fluid will spray out under pressure from the unit. It is advised that eye protection be worn during removal of the seal and care taken to protect footwear and clothing.

7   Place the suspension unit upright with its end in a drip tray or similar container and wrap a piece of rag around the base of the unit casing. This rag will prevent most of the ATF from spraying out sideways as the seal leaves the casing. Remove the seal by pushing the compressed air hose against the air hose connection at the top of the unit and applying compressed air in short bursts to blow the seal from the unit. Remove the backing ring end of the unit on the worksurface. It is not possible to proceed any further with dismantling of the unit.

8   Inspect the bush for wear. Honda supply no figures for the amount of wear allowed on the bearing surfaces of the bush; it is therefore a matter of judgement and comparison whether the bush needs to be renewed. If in doubt, return the bush to an official Honda Service Agent who will be able to supply a new item for comparison and give his opinion on the matter.

9   Clean all the component parts and lay them out on a clean work surface ready for reassembly. Inspect the surface of the tube, where it passes through the seal, for scoring or damage to the surface caused by dirt or moisture finding its way through the protective gaiter. Any damage to the tube finish will cause leakage of ATF past the oil seal and also effectively reduce the life of any new seal fitted. It is therefore considered advisable to renew the unit if this sort of damage has occurred.

10   Inspect the rubber inserts located in the mounting lugs at each end of the unit. These may be pushed out for inspection and if found to be worn or perished, replaced with new items. Finally, discard the oil seal and replace it with a new item.

11   Plug the air hose connection, invert the unit and refill it with 365 cc (12.34/12.85 US/Imp fl oz) of ATF. Commence reassembly by inserting the bush followed by the backing ring into the casing. Ensure that the bush is fitted the corect way up with the lips facing the air hose connection of the unit.

12   Lubricate the new oil seal with ATF and carefully press it into the casing with fingers or thumbs until its upper surface is just below the casing edge. Fit the seal with the lip facing the air hose connection point. Place the backing ring over the seal and gently drift both seal and ring further into the casing until the circlip locating groove is exposed. It will be necessary to manufacture a tool out of a length of thick-walled tubing, with which to locate the seal and backing ring. The internal diameter of the tube should be just large enough to allow it to pass over the tube of the unit and locate on the backing ring. Ensure that the end of the tube is square to its sides and free from burrs. Using this tool to drift in the seal and ring will ensure that both components remain square to the casing during fitting. Alternatively, Honda supply tool no 07947 - 4630100 for this purpose.

13   Refit the circlip with the chamfered edge facing the backing ring. Refit the rubber gaiter. Smear the new O-ring for the air hose connection with grease before fitting the air hose to the unit and tightening the connection to a torque of 0.4 - 0.7 kgf m (3 - 5 lbf ft). Note that a special 'crowsfoot' adaptor will be needed to enable a torque wrench to be used on the hose connection. If this tool is not available, the connection should be screwed in until finger-tight and then nipped tight with an open-ended spanner. Do not overtighten the connection; there is a high risk of its shearing.

14   Refitting the units is a reversal of the removal procedure, noting the following points. After connecting the two air hoses to the charging union, torque load them to 1.5 - 2.0 kgf m (11 - 14 lbf ft) using the method described in the previous paragraph. If the air pressure sensor has been removed from the union for any reason, a new O-ring should be greased and fitted to the sensor before the sensor is fitted to the union and tightened to a torque of 0.8 - 1.2 kgf m (6 - 9 lbf ft).

15   Refit the suspension units to the machine. Some difficulty may be experienced when refitting the suspension units insomuch as the rubber inserts in each unit mounting lug will easily become displaced if the holes in the inserts are not aligned exactly with the mounting points on the machine before pushing the unit into position. Note the correct location of the mounting washers by referring to the accompanying figure. All suspension unit mounting nuts (and bolt) should be tightened to a torque of 3.0 - 4.0 kgf m (22 - 29 lbf ft).

16   Ensure both air hoses are rotated correctly and reinserted in the clamp on the frame and fuel tank. It is important that the hoses are not allowed to chafe on any of the frame components. Refit the charging union and recharge both suspension units to within the range given in the Specifications, with the machine placed on its centre stand so that the weight of the machine is taken off the rear suspension. Use a method of charging similar to that given in the Section 3 for the front forks. On completion of charging, refit and tighten the chromed protector cap to the Shrader valve. Check that the air pressure sensor electrical connection points are clean before reconnecting them and reclipping the wire to the mounting plate.

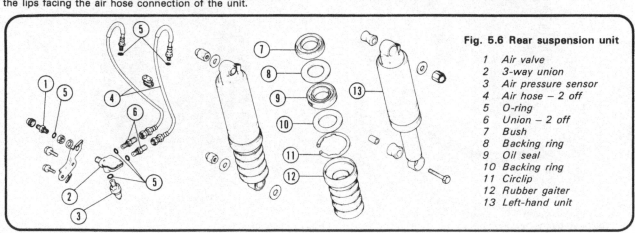

Fig. 5.6 Rear suspension unit

1   Air valve
2   3-way union
3   Air pressure sensor
4   Air hose – 2 off
5   O-ring
6   Union – 2 off
7   Bush
8   Backing ring
9   Oil seal
10   Backing ring
11   Circlip
12   Rubber gaiter
13   Left-hand unit

10.8 Inspect the bush for wear

10.10 The rubber inserts may be pushed out of the mounting lugs

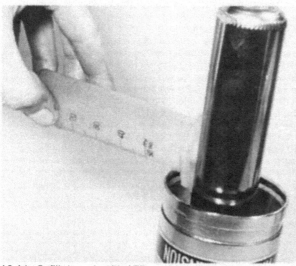

10.11a Refill the unit with ATF ...

10.11b ... before inserting the bush ...

10.11c ... followed by the backing ring ...

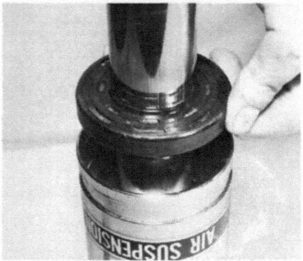

10.12 ... the new oil seal ...

10.13 ... the second backing ring and the retaining circlip

10.16a Recharge both units to the correct pressure

10.16b Connect the pressure sensor leads and clip them to the mounting plate

11 Centre stand: examination

1   The centre stand pivots on a hollow tube, clamped between two split lugs below the rear engine mounting. The pinch bolts and split-pin should be periodically checked for security and the pivot oiled.

2   Check also that the return spring(s) have not stretched and that the linkage is unworn; the stand must retract smartly and be firmly held in the retracted position by the spring(s). If the stand falls whilst the machine is in motion, it may well catch on the road surface causing the rider to be thrown from the machine.

12 Prop stand: examination

1   The prop stand pivots on a frame lug located beneath the lower mid-point engine mounting on the left-hand side of the machine. Check the pivot bolt and nut for security and signs of wear and oil the pivot assembly occasionally. Check that the return spring has not stretched and that it returns the stand to the retracted position smartly and holds it in place. An accident will almost inevitably result from the prop stand dropping whilst the machine is in motion.

2   To prevent the machine from being ridden away with the prop stand down, the GL 1100 model incorporates a self-retracting device. Check the rubber 'trip' of this device for wear or damage. No part should be worn below the moulded line on the rubber.

3   Check the operation of the stand as follows: Put the machine on the centre stand, and put the prop stand down. Using a spring balance attached to the extreme end of the centre stand, measure the force required to retract the stand. If this force exceeds 1.5 - 2.5 kg (3.3 - 5.5 lb), check that the stand pivot is not overtight, or in need of lubrication.

4   To renew the rubber 'trip', take off the bolt. Make sure the sleeve is installed in the fixing hole of the new trip. Fit the trip with the arrow facing outwards. The block should bear the marking 'over 260 lbs only'.

13 Footrests: examination

1   The front footrests are retained on the frame by a single bolt and comprise a spring loaded rubber covered peg attached by a clevis pin to a forged arm. The footrest rubber is held to the support arm by a bolt passing through a plate below the rubber.

2   Because the footrest pegs are spring loaded and can fold back when struck, it is unlikely that they will become damaged. The support arms can be repaired, if bent, by the application of heat before being straightened.

3   To renew a return spring, remove the split pin and pull out the clevis pin from the footrest pivot piece.

4   The pillion footrests are also of the folding type, spring loading being provided by the compression of the footrest rubber. Owing to the construction, it is unlikely that these footrests can be repaired, if damaged.

14 Rear brake pedal: examination

1   The rear brake pedal is mounted on, and pivots on, a shaft fixed inboard of the frame, just below the swinging arm pivot stub on the right-hand side of the machine. The pedal is retained by a washer and split pin. The pedal return spring is fitted on the pivot shaft between the pedal and washer.

2   The brake pedal is connected directly to the piston rod of rear brake master cylinder by means of a forked joint and clevis pin. The piston rod and fork are threaded, to allow height adjustment of the pedal.

3   If the pedal becomes bent during an accident, it may be straightened, after removal from the machine. Apply heat to the

11.1 Check the centre stand pivot assembly for security

14.2 The brake pedal and master cylinder assembly

damaged area with a blowtorch or gas bottle. Bear in mind that the required amount of heat will probably cause the chrome plating to peel from the pedal. This is unavoidable.

15 Dualseat: removal and refitting

1   The dualseat is retained at the rear by two Allen bolts passing through brackets attached to the seat pan. The seat front is supported on a rubber saddle which locates on a pressed steel bridge across the frame tubes.
2   Removal of the seat is achieved by unscrewing the Allen bolts, moving the seat backwards to detach it from the front mounting and lifting the front of the seat clear of the rear of the dummy fuel tank so that it can be pulled forwards and away from the machine. Refitting the seat can be carried out by reversing the above procedure.

16 Dummy fuel tank: removal and refitting

1   The dummy fuel tank is fitted in place of a receptacle for fuel and acts as a protective cover for the air filter assembly, the radiator tank, the electrical components mounted on the frame beneath it and the fuel filler point. It also contains the machine's tool tray.
2   The unit may be removed from the machine by first opening the flaps on its upper surface and lifting out the tool tray. Remove the dualseat and detach the tank from its frame mounting points by unscrewing the four securing bolts, two at the front of the tank and two at the rear. The tank may now be carefully lifted up away from the machine, bearing it slightly to the right so that it clears the components located within it. Refitting the unit can be carried out by reversing the above procedure.

17 Instrument heads: removal and refitting

1   To remove the speedometer or tachometer heads, disconnect the drive cable from the instrument by unscrewing the knurled cable nut. Unscrew and remove the two acorn nuts and washers, disconnect the head from the rubber sleeve and baseplate and pull out the bulb holders from the instrument base. Remove the head. Refitting the instrument heads is a reversal of the above procedure.
2   It is not possible to repair a faulty instrument head. If the instrument fails completely, or moves jerkily, first check the

drive cable. If the mileage recorder of the speedometer ceases to function but the speedometer continues to work or vice versa, the instrument head is faulty.
3   Remember that a working speedometer, accurate to within 10% is a statutory requirement in the UK and many other countries.

18 Instrument drive cables: removal and fitting

1   The speedometer and tachometer drive cables should be examined and lubricated occasionally. The outer sheath should be examined for cracks or damage, the inner cable for broken or frayed strands. Jerky or sluggish instrument movement is generally caused by a faulty cable.
2   Detach the cable at the drive end, and withdraw the inner cable. Clean and examine the cable. Re-lubricate it with high melting point grease, but do not grease the top six inches of cable, at the instrument end, or grease will work its way into the instrument head and ruin the movement. On some models removal of the inner cable is not possible because of the design. In these cases lubrication *can* be accomplished using a hydraulic lubricator and oil. This procedure is not recommended, however because of the likelihood of the relatively low viscosity oil finding its way into the instrument head.
3   Re-route the cables as they were originally. Make sure that the steering turns freely.

19 Fairing and mounting bracket: removal and refitting – Interstate model

1   Commence removal of the fairing assembly by unscrewing the three screws that retain each lower side fairing panel in position. Lift away each panel in turn and place it, inside face downwards, on a clean protected surface. It is imperative that all fairing panels are carefully stored by this method so that damage to the outer surface finish is avoided.
2   Trace the wiring harness to the 9-pin block connector situated on the left-hand side of the fairing and separate the two halves of the connector. With an assistant supporting the fairing, unscrew and remove the four nuts, plain washers, lock washers and flange bolts that retain the fairing to the mounting bracket; the fairing can now be lifted from position.
3   Before removal of the fairing mounting bracket assembly can take place, the seat and dummy fuel tank must be removed by using the procedures given in the Sections 15 and 16 of this Chapter. Remove the horns from their mounting brackets by unplugging the electrical connections from each horn body and

unscrewing and removing each retaining bolt.

4 Detach the left- and right-hand end brackets from the main part of the mounting assembly by removing the two 14 mm flange bolts that retain each bracket in position. Detach the left- and right-hand inner covers by removing their retaining screws.

5 Refer to the accompanying figure and remove the crash bars protecting each cylinder head. Remove the single 8 mm bolt the retains the mounting bracket to a point just behind the steering head. Remove the remaining four 6 mm nuts and gripping the front of the mounting bracket, ease it away from the frame.

6 Refitting of the mounting bracket assembly and fairing is a reversal of the procedure given for removal. Refer to the torque figures given in the Specifications of this Chapter when tightening the retaining bolts and nuts.

20 Windshield: adjustment – Interstate model

1 The height of the windshield on the fairing of an Interstate model may be adjusted 25 mm (1.0 m) in either direction of the mid-position. To adjust the windshield height, loosen the screws retaining the rearview mirrors and trim, and move the windshield to the required height. Return it in position by tightening the screws in the reverse order to that shown.

21 Headlamp: removal and refitting – Interstate models

1 To remove the headlamp assembly from its location in the fairing of an Interstate model, first remove the adjusting knob from within the fairing by loosening its set screw and pulling it off its shaft. Unscrew and remove the nut from the threaded sleeve, followed by the lockwasher and plain washer. The headlamp can now be eased out of the fairing. Reverse the above procedure to refit the headlamp.

22 Cleaning the machine

1 After removing all surface dirt with a rag or sponge which is washed frequently in clean water, the machine should be allowed to dry thoroughly. Application of car polish or wax to the cycle parts will give a good finish, particularly if the machine receives this attention at regular intervals.

2 The plated parts should require only a wipe with a damp rag, but if they are badly corroded, as may occur during the winter when the roads are salted, it is permissible to use one of the proprietary chrome cleaners. These often have an oily base which will help to prevent corrosion from recurring.

4 If the engine parts are particularly oily, use a cleaning compound such as Gunk or Jizer. Apply a compound whilst the parts are dry and work it in with a brush so that it has an opportunity to penetrate and soak into the film of oil and grease. Finish off by washing down liberally, taking care that water does not enter the carburettors, air cleaners or the electrics.

5 If possible, the machine should be wiped down immediately after it has been used in the wet, so that it is not garaged under damp conditions that will promote rusting.

6 Remember there is less chance of water entering the control cables and causing stiffness if they are lubricated regularly as described in the Routine Maintenance section.

23 Fault diagnosis: frame and forks

Symptom	Cause	Remedy
Machine veers to left or right with hands off handlebars	Wheels out of alignment Forks twisted Frame bent	Check wheels and realign Strip and repair Strip and repair or renew
Machine tends to roll at low speeds	Steering head bearings not adjusted correctly or worn	Check adjustments and renew bearings, if worn
Machine tends to wander	Worn swinging arm bearings	Check and renew bearings Check adjustment and renew
Forks judder when front brake is applied	Steering head bearings slack Forks worn on sliding surfaces	Dismantle, lubricate and adjust Dismantle and renew worn parts
Forks bottom	Short of oil Insufficient air pressure	Replenish with correct viscosity oil Check air pressure and recharge
Fork action stiff	Fork legs out of alignment Bent shafts, or twisted yokes Too much air pressure	Strip and renew, or slacken clamp bolts, front wheel spindle and top bolts. Pump forks several times, and tighten from bottom upwards Release pressure until gauge reading is within limits
Machine tends to pitch badly	Defective rear suspension units, or ineffective fork damping Front forks or rear suspension units incorrectly charged	Check damping action Check the grade and quantity of oil in the front forks Check units are charged to within given limits

Chapter 6 Wheels, brakes and tyres

Contents

Specifications

Wheels

Type	Comstar

Brakes

Front:

Type	Hydraulically operated twin disc
Disc thickness	4.9 – 5.1 mm (0.19 – 0.20 in)
Service limit	4.0 mm (0.16 in)
Disc runout (max)	0.3 mm (0.01 in)
Master cylinder ID	15.870 – 15.913 mm (0.6248 – 0.6265 in)
Service limit	15.925 mm (0.6270 in)
Master cylinder piston OD	15.827 – 15.854 mm (0.6231 – 0.6242 in)
Service limit	15.815 mm (0.6226 in)
Caliper cylinder ID	42.850 – 42.900 mm (1.6870 – 1.6890 in)
Service limit	42.920 mm (1.6898 in)
Caliper piston OD	42.772 – 42.822 mm (1.6839 – 1.6859 in)
Service limit	42.765 mm (1.6837 in)
Pad thickness	6.9 – 7.1 mm (0.27 – 0.28 in)
Service limit	Red line

Rear:

Type	Hydraulically operated single disc
Disc thickness	6.9 – 7.1 mm (0.27 – 0.28 in)
Service limit	6.0 mm (0.24 in)
Disc runout (max)	0.3 mm (0.01 in)
Master cylinder ID	14.0 – 14.043 mm (0.5512 – 0.5529 in)
Service limit	14.055 mm (0.5533 in)
Master cylinder piston OD	13.957 – 13.984 mm (0.5495 – 0.5506 in)
Service limit	13.940 mm (0.5488 in)
Caliper cylinder ID	42.850 – 42.900 mm (1.6870 – 1.6890 in)
Service limit	42.920 mm (1.6898 in)
Caliper piston OD	42.772 – 42.822 mm (1.6839 – 1.6859 in)
Service limit	42.765 mm (1.6837 in)
Pad thickness	7.9 – 8.1 mm (0.31 – 0.32 in)
Service limit	Red line
Fluid specification	DOT 3 (USA) or SAE J1703 (UK)

Tyres

Size:
Front 110/90-19 62H
Rear 130/90-17 68H
Pressure:
Front and rear* .... 32 psi (2.25 kg/cm²)
*Increase rear tyre pressure to 40 psi (2.80 kg/cm²) when carrying a load greater than 200 lb (90 kg)

Final drive bevel housing

Oil capacity .... 140 – 160 cc (5.07/5.28 US/Imp fl oz)

Torque wrench settings

	kgf m	lbf ft
Front wheel spindle clamp nuts (10 mm)	3.0 – 4.0	22 – 29
Front wheel spindle securing nut (12 mm)	5.5 – 6.0	40 – 43
Rear wheel spindle securing nut (18 mm)	8.0 – 10.0	58 – 72
Rear wheel spindle pinch bolt (8 mm)	2.4 – 2.9	17 – 21
Front brake disc securing bolt nuts (8 mm)	2.7 – 3.3	20 – 24
Front brake caliper mounting bolts (10 mm)	3.0 – 4.0	22 – 29
Front brake caliper bolts (8 mm)	1.5 – 2.0	11 – 14
Front brake master cylinder securing clamp bolts (6 mm)	0.8 – 1.2	6 – 9
Front brake hose 2-way joint bolts (6 mm)	1.0 – 1.4	7 – 10
Front brake hose union bolts (10 mm)	2.5 – 3.5	18 – 25
Rear brake disc securing nuts (8 mm)	2.7 – 3.3	20 – 24
Rear brake caliper torque arm securing nut (8 mm)	1.8 – 2.5	13 – 18
Rear brake master cylinder securing bolt (8 mm)	2.4 – 2.9	17 – 21
Rear brake fluid reservoir securing bolt (6 mm)	1.0 – 1.4	7 – 10
Rear brake caliper bolts (8 mm)	1.5 – 2.0	11 – 14
Rear brake hose union bolts (10 mm)	2.5 – 3.5	18 – 25
Rear brake rod joint locknut (8 mm)	1.5 – 2.0	11 – 14
Final drive bevel housing filler cap (30 mm)	1.0 – 1.4	7 – 10
Final drive bevel housing drain bolt (6 mm)	1.0 – 1.4	7 – 10

1 General description

The Honda GL 1100 is fitted as standard with Honda Comstar wheels. Each wheel has an aluminium alloy hub and rim interconnected by ten pressed-steel or aluminium alloy spoke blades. The blades are riveted to the rim and secured to the hub by through bolts. Although the wheels are fabricated in this manner, once manufactured, they are considered as single assemblies. Both wheels are fitted with tubeless tyres of which the front tyre size is 110/90-19 62H and the rear tyre size 130/90-17 68H. The 'H' speed rating of these tyres denotes that they are safe up to speeds of 130 mph.

The front wheel is fitted with twin hydraulically-operated disc brakes, the calipers being of the self-aligning type, using a single piston and located to the rear of the fork lower legs. The front brake master cylinder and reservoir is mounted in an exposed position on the handlebar, where it is operated directly by the lever.

The rear wheel is fitted with a single hydraulically-operated disc brake of the same type as fitted to the front; the brake being operated by a foot pedal on the right-hand side of the machine and having a reservoir and master cylinder located beneath the right-hand sidepanel.

Final drive is provided by a shaft running through the right-hand member of the swinging arm, to which is attached a crown and pinion final drive assembly contained within an alloy housing. A cush drive assembly is interposed between the final drive gear and the rear wheel, to absorb shocks in the transmission.

2 Front wheel: examination

1 Place the machine on its centre stand, with a block under the sump, so that the front wheel is clear of the ground.
2 Spin the wheel and check for rim alignment by placing a pointer close to the rim edge. If the total radial or axial alignment variation is greater than 2.0 mm (0.08 in) the manufacturer recommends that the wheel is renewed. This policy, is however, a counsel of perfection and in practice a larger runout may not affect the handling properties excessively.
3 Although Honda do not offer any form of wheel rebuilding facility, a number of private engineering firms offer this service. It should be noted however, that Honda do not approve of this course of action.
4 Check the rim for localised damage in the form of dents or cracks. The existence of even a small crack renders the wheel unfit for further use unless it is found that a permanent repair is possible using arc-welding. This method of repair is highly specialised and therefore the advice of a wheel repair specialist should be sought.
5 Because tubeless tyres are used, dents may prevent complete sealing between the rim and tyre bead. This may not be immediately obvious until the tyre strikes a severely irregular surface, when the unsupported tyre wall may be deflected away from the rim, causing rapid deflation of the tyre. Honda recommend that the wheel be renewed if the bead seating surface of the rim is scratched to a depth of 0.5 mm (0.02 in) or more. Again, if in doubt, seek specialist advice whether continued use of the wheel is advisable.
6 Inspect the spoke blades for cracking and security. Check carefully the area immediately around the rivets which pass through the spokes and into the rim. In certain circumstances where steel spokes are fitted electrolytic corrosion may occur between the spokes, rivets and rim due to the use of different metals.

3 Front wheel: removal and refitting

1 With the front wheel supported well clear of the ground, remove the cross head screw which retains the speedometer cable to its drive gearbox, on the left-hand side of the hub. Pull the cable out and refit the screw, to prevent loss.
2 Remove the two bolts holding one of the caliper support brackets to the fork leg and lift the caliper and bracket assembly

off the disc. Support the weight of the caliper with a length of string or wire attached to the frame or engine.

3   Slacken and remove the clamp nuts at the base of each fork leg. With the clamps released, the wheel will drop free and can be manoeuvred clear of the forks and mudguard.

4   Do not operate the front brake lever while the wheel is removed since fluid pressure may displace the pistons and cause leakage. Additionally, the distance between the pads will be reduced, making refitting of the brake discs more difficult.

5   Refit the wheel by reversing the dismantling procedure. Do not omit the spacer which is a push fit in the oil seal on the right-hand side of the wheel or the speedometer gearbox which is a push fit on the left-hand side. Ensure that the speedometer drive dogs engage with the notches in the gearbox drive sleeve. Lift the wheel into position whilst ensuring that the speedometer gearbox is positioned correctly. Fit the clamps to the base of each fork leg; the arrows marked on each clamp must face forward. Fit the clamp retaining nuts and washers finger-tight.

6   Carefully lower the disconnected brake caliper assembly over the disc so as to avoid damage to the pads and fit and tighten the two securing bolts to a torque of 3.0 – 4.0 kgf m (22 – 29 lbf ft). Tighten the nuts of the left-hand clamp only to a torque of 3.0 – 4.0 kgf m (22 – 29 lbf ft), starting with the forward nut.

7   Before tightening the right-hand clamp securing nuts, it is first necessary to determine that the clearance between the outside surface of the right-hand disc and the rear of the caliper support bracket is correct. If this clearance is not correct, damage to the disc is likely to occur resulting in impaired braking efficiency or worse. With a feeler gauge of 0.7 mm (0.028 in) thickness, measure the clearance. If the gauge inserts easily into the space between the disc and support bracket, torque load the clamp securing nuts to 3.0 – 4.0 kgf m (22 – 29lbf ft), starting with the forward nut.

8   If the feeler gauge will not insert into the space, grasp the right-hand fork lower leg and pull it outwards until the gauge can be inserted. Tighten the clamp securing nuts as stated above and withdraw the feeler gauge. Check that the clearance between the disc surface and the other three corners of the caliper support bracket is also 0.7 mm (0.028 in). Spin the wheel to ensure that it revolves freely and check the brake operation. Check that all nuts and bolts are fully tightened. If the clearance between the disc and pads is incorrect pump the front brake lever several times to adjust. Finally, reconnect the speedometer cable to the speedometer gearbox connection.

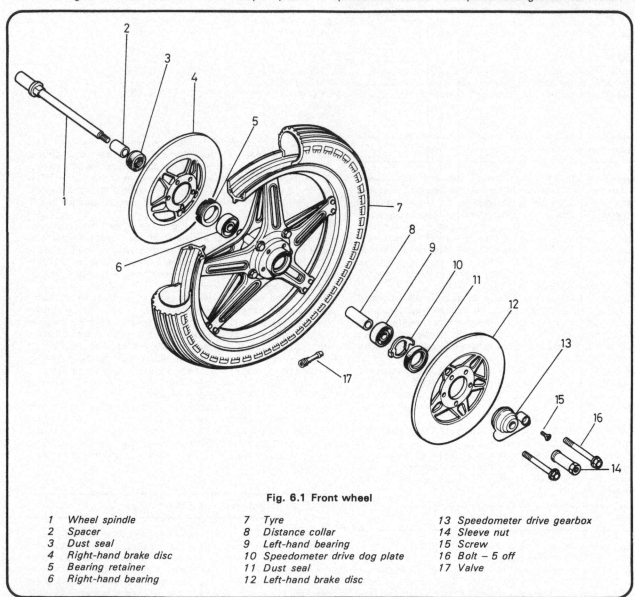

Fig. 6.1 Front wheel

1	Wheel spindle	7	Tyre	13	Speedometer drive gearbox
2	Spacer	8	Distance collar	14	Sleeve nut
3	Dust seal	9	Left-hand bearing	15	Screw
4	Right-hand brake disc	10	Speedometer drive dog plate	16	Bolt – 5 off
5	Bearing retainer	11	Dust seal	17	Valve
6	Right-hand bearing	12	Left-hand brake disc		

3.1 Remove the retaining screw to release the speedometer cable

3.2 Remove the two bolts which retain the caliper support to the fork leg

3.3 Remove the clamp from the base of each fork leg

3.5 Ensure the speedometer gearbox is positioned correctly

4   Front wheel bearings: removal, examination and fitting

1   With the wheel removed from the machine, as described in the previous Section, unscrew and remove the sleeve nut from the wheel spindle. Pull the speedometer gearbox from its location in the left-hand side of the wheel hub and remove the wheel spindle, followed by the spacer contained within the right-hand dust seal. The wheel bearings are of the ball journal type and non-adjustable. There are two bearings and two dust seals, the two bearing being separated by a distance collar in the centre of the hub. To avoid damage occurring to the brake discs during removal of the bearings, it is recommended that they be removed by following the procedure given in Section 7.

2   Note that a threaded retainer is fitted to the right-hand side of the hub, and this must be removed before the bearings can be released. A suitable peg spanner should be fabricated unless the Honda tool, No 07710-0010200, is available. Do not resort to using a punch to loosen the retainer, this will only result in damage. The retainer will be staked in position and will require firm, even pressure to release it. Note also that new bearings and seals should be fitted whenever the old items are removed, so check carefully for wear before dismantling commences.

3   The left-hand bearing should be drifted out first, from the right-hand side of the wheel. Use a long drift against the inner face of the inner race. It may be necessary to knock the collar to one side so that purchase can be made against the race. Work round in a circle to keep the bearing square in the housing. The dust seal and speedometer drive dog plate will be pushed out as the bearing is displaced. After removal of the bearing, take out the distance collar and then drift out the right-hand bearing and dust seal in a similar manner, from inside the hub.

4   Remove all the old grease from the hub and bearings, wash the bearings in petrol and dry them thoroughly. Check the bearings for roughness by spinning them whilst holding the inner race with one hand and rotating the outer track with the other. If there is the slightest sign of roughness renew them.

5   Before driving the bearings back into the hub, pack the hub with new grease and also grease the bearings. Use a tubular drift of the same diameter as the outer race of each bearing to drive the bearings back into the wheel hub. Ensure that the bearings enter the hub squarely and are fitted with the closed side facing outwards. Do not omit to refit the distance collar before fitting the second bearing.

6   Locate the speedometer drive plate in the wheel hub, ensuring that the tab of the plate fits in the slot in the hub casting. Fit the new dust seal over the drive plate. Screw the

bearing retainer into position, having obtained a new replacement if there were signs of wear or damage to the threads. Tighten the retainer firmly, then secure it in this position by staking at the junction of the retainer and the wheel hub. A locking compound may be used to secure the ring in preference to staking. Refit the brake discs, tightening the retaining bolt nuts evenly and in a diagonal sequence to a torque of 2.7 – 3.3 kgf m (20 – 24 lbf ft).

7   Refit the spacer into the right-hand dust seal and push the wheel spindle into position in the wheel after first having smeared its bearing surfaces lightly with grease. Lubricate the speedometer gearbox with the correct type of grease and push it into position, aligning the tangs on the gearbox with the notches in the drive plate. Fit and tighten the wheel spindle sleeve nut to a torque of 5.5 – 6.0 kgf m (40 – 43 lbf ft).

5   Front disc brake assembly: examination and brake pad renewal

1   Check the front brake master cylinder, hoses and caliper units for signs of leakage. Pay particular attention to the condition of the hoses, which should be renewed without question if there are signs of cracking, splitting or other exterior damage. Check the hydraulic fluid level by referring to the upper and lower level lines visible on the exterior of the translucent reservoir body.

2   Replenish the reservoir after removing the cap on the brake fluid reservoir and lifting out the diaphragm plate. The reservoir cap is attached to the reservoir base by four securing screws. The condition of the fluid is one of the maintenance tasks which should **never be neglected.** If the fluid is below the level mark, brake fluid of the correct specification must be added. **Never** use engine oil or any fluid other than that recommended. Other fluids have unsatisfactory characteristics and will rapidly destroy the seals.

3   The two sets of brake pads should be inspected for wear. Each had a red groove, which marks the wear limit of the friction material. When this limit is reached, both pads in the set must be renewed, even if only one has reached the wear mark. A small inspection window, closed by a plastic cap, is provided in the top of each caliper unit so that examination of pad condition may be carried out easily.

4   If the brake action becomes spongy, or if any part of the hydraulic system is dismantled (such as when a hose has been renewed) it is necessary to bleed the system in order to remove all traces of air. Follow the procedure in Section 9 of this Chapter.

5   To gain access to the pads for renewal, the caliper assembly being attended to must be partially dismantled as follows; removal of the wheel is not required. Unscrew the two bolts that pass into the caliper casing. Carefully ease the casing up leaving the pads in position either side of the disc and supported by the caliper mounting bracket. The pads can be lifted from place, one at a time.

6   Refit the new pads and refit the caliper by reversing the dismantling procedure. The caliper piston should be pushed inwards slightly so that there is sufficient clearance between the brake pads to allow the caliper to fit over the disc. Do not omit the anti-chatter shim which should be fitted on the rear face of the piston side pad with the arrow pointing forward (facing the direction of wheel rotation). It is recommended that the outer periphery of the outer (piston) pad is lightly coated with disc brake assembly grease (silicone grease). Use the grease sparingly and ensure that grease **does not** come into contact with the friction surface of the pad.

7   Before refitting the caliper casing, check the condition of the dust covers around the slider spindles and of the spindles themselves. Check also the condition of the caliper piston seal and boot. If the condition of any of these components is seen to be doubtful, refer to the following Section for further information. With the caliper casing refitted, fit and tighten the two securing bolts to a torque of 1.5 – 2.0 kgf m (11 – 14 lbf ft).

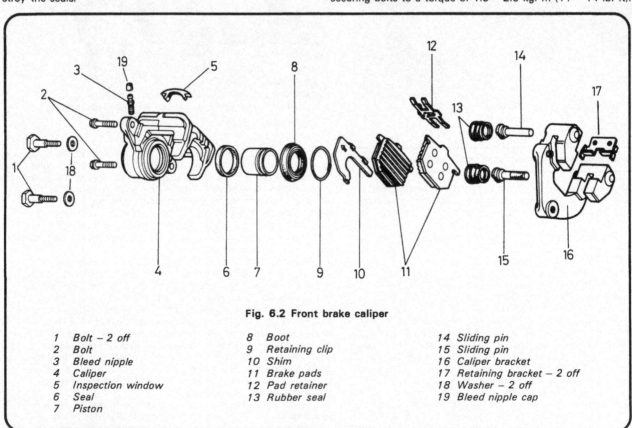

Fig. 6.2 Front brake caliper

1 Bolt – 2 off	8 Boot
2 Bolt	9 Retaining clip
3 Bleed nipple	10 Shim
4 Caliper	11 Brake pads
5 Inspection window	12 Pad retainer
6 Seal	13 Rubber seal
7 Piston	

14   Sliding pin
15   Sliding pin
16   Caliper bracket
17   Retaining bracket – 2 off
18   Washer – 2 off
19   Bleed nipple cap

4.6a Locate the speedometer drive plate in the wheel hub ...

4.6b ... followed by the new dust seal

4.7a Push the new dust seal into position ...

4.7b ... insert the spacer through the seal ...

4.7c ... and insert the spindle through the wheel hub

4.7d Push the speedometer gearbox into position over the spindle

5.2 The brake fluid reservoir cap and diaphragm assembly is retained by four screws

5.5a Detach the caliper from its mounting bracket ...

5.5b ... to gain access to the brake pads

5.6 Refit the shim with the arrow pointing in the direction of wheel rotation

5.7 Check the condition of the slider spindle dust covers

6 Front brake calipers: examination and overhaul

1   It is recommended that the two caliper units are removed and overhauled separately, to prevent the accidental transposition of identical components. Note that any work on the hydraulic system must be undertaken under ultra-clean conditions. Particles of dirt will score the working parts and cause early failure. Select a suitable receptable into which may be drained the hydraulic fluid. Remove the banjo bolt holding the hydraulic hose at the caliper and allow the fluid to drain. Take great care not to allow hydraulic fluid to spill onto paintwork; it is a very effective paint stripper. Hydraulic fluid will also damage rubber and plastic components.
2   Remove the caliper from the fork leg and displace the brake pads as described in the preceding Section. Withdraw the two slider spindles and rubber boots from the support bracket.
3   Displace the circlip which holds the piston boot in position and then prise out the piston boot, using a small screwdriver, taking care not to scratch the surface of the cylinder bore. The piston can be displaced most easily by applying an air jet to the hydraulic fluid feed orifice. Be prepared to catch the piston as it falls free. Displace the annular piston seal from the cylinder bore groove, again using the flat of a small screwdriver and taking

care not to scratch the surface of the cylinder bore.

4   Clean the caliper components thoroughly in trichlorethylene or in hydraulic brake fluid. **CAUTION:** Never use petrol for cleaning hydraulic brake parts otherwise the rubber components will be damaged. Discard all the rubber components as a matter of course. The replacement cost is relatively small and does not warrant re-use of components vital to safety.

5   Check the piston and caliper cylinder bore for scoring, rusting or pitting. If any of these defects are evident it is unlikely that a good fluid seal can be maintained and for this reason the components should be renewed. Refer to the service limits given in the Specifications for the overall diameter of the piston and the internal diameter of the caliper bore. Measure the two components and renew if worn beyond these limits. Inspect the slider spindles for wear and check their fit in the support bracket. Slack between the spindles and bores may cause brake judder if wear is severe.

6   To assemble the caliper, reverse the removal procedure. When assembling pay attention to the following points. Apply caliper grease (high heat resistant) to the caliper spindles. Apply a generous amount of brake fluid to the inner surface of the cylinder and to the periphery of the piston, then reassemble. Do not reassemble the piston with it inclined or twisted. When installing the piston push it slowly into the cylinder while taking care not to damage the piston seal. Apply brake pad grease around the periphery of the moving pad.

7   Refer to Section 9 and bleed the brake system after refilling the reservoir with new hydraulic fluid of the correct specification. Check for leakage whilst applyiing the brake lever tightly and repeat the entire servicing procedure for the second brake caliper. Test the operation of the brakes by pushing the machine forward and applying the brake lever. If this test is satisfactory, test run the machine applying the brakes soon after riding away and at intervals thereafter and noting the level of the hydraulic fluid in the handlebar mounted reservoir to ensure it does not drop. On completion of the test run, recheck the system for signs of leakage and check the disturbed connections for security.

7   Front brake discs: examination, removal and fitting

1   It is unlikely that either of the two discs will require attention unless they become badly worn or scored or in the unlikely event of warpage.

2   Warpage should be measured with the discs still attached to the wheel and the wheel in situ on the machine using a dial gauge. The maximum permissible warpage for both discs is 0.3 mm (0.012 in).

3   Disc removal is straightforward after the wheel has been

7.4 The wear service limit is marked on each brake disc

taken out from the machine as described in Section 3 of this Chapter. The two discs are retained on the wheel hub by five flange bolts and nuts which pass through the left-hand disc centre plate, the hub casting and the right-hand disc centre plate. The retaining nuts are fitted on the right-hand disc and torque loaded to 2.7 – 3.3 kgf m (20 – 24 lbf ft). Undo these nuts evenly and in a diagonal sequence and tighten them using the same technique.

4   The brake discs will wear eventually to a thickness which no longer provides sufficient support, and will probably begin to warp. The correct wear service limit, which can be measured with a micrometer is 4.0 mm (0.16 in).

8   Front brake master cylinder: removal, examination, renovation and refitting

1   The master cylinder and hydraulic reservoir take the form of a combined unit mounted on the right-hand side of the handlebars, to which the front brake lever is attached. The master cylinder is actuated by the front brake lever, and applies hydraulic pressure through the system to operate the front brakes when the handlebar lever is manipulated. The master cylinder pressurises the hydraulic fluid in the brake pipe which, being incompressible, causes the piston to move in the caliper unit and apply the friction pads to the brake disc. If the master cylinder seals leak, hydraulic pressure will be lost and the braking action rendered much less effective.

2   Before the master cylinder can be removed, the system must be drained. Place a clean container below one caliper unit and attach a plastic tube from the bleed screw on top of the caliper unit to the container. Open the bleed screw one complete turn and drain the system by operating the brake lever until the master cylinder reservoir is empty. Close the bleed screw and remove the pipe. Remember that brake fluid is a good paint stripper; do not allow it to come into contact with painted surfaces or plastic components.

3   Disconnect the front brake stop lamp switch wire at the push connector. Unscrew the union bolt and disconnect the connection between the master cylinder and brake hose. Tie the hose to a point on the fork assembly and mask its end to prevent the ingress of dirt and moisture into the brake system. Remove the rear view mirror by unscrewing it from the master cylinder body. Unscrew the brake lever retaining nut and bolt and detach the lever from its pivot lugs. Unscrew the two master cylinder fastening bolts and remove the master cylinder body from the handlebars. Empty any surplus fluid from the reservoir.

4   Remove the circlip located beneath the cylinder body boot, followed by the plain washer, piston (with secondary cup), primary cup and the spring. Note that it may be necessary to apply a low pressure air supply to the master cylinder outlet in order to displace the piston. Place the components in a clean container and wash them in new brake fluid. Examine the cylinder bore and piston for scoring. Renew if scored. Examine the brake lever pivot point and the master cylinder pivot lugs for wear, cracks or fractures, and the hose union threads and brake pipe threads for cracks or other signs of deterioration. Remove the reservoir from the cylinder body by unscrewing and removing the four securing screws. Renew the O-ring attached to the base of the reservoir and inspect the diaphragm for signs of damage or deterioration, renewing it if necessary.

5   When reassembling the master cylinder follow the removal procedure in reverse order. Renew the various seals, lubricating them with silicone grease or hydraulic fluid before they are refitted. Make sure that the primary and secondary cups are fitted the correct way round and that their lips do not turn inside out when being fitted. Check that, once fitted, the circlip is correctly located in its retaining groove.

6   Mount the master cylinder on the handlebars so that the fluid reservoir is horizontal when the motorcycle is on the centre stand with the steering in the straight ahead direction. Note that Honda mark the handlebars and master cylinder securing

bracket to ensure correct positioning of the unit. The securing bracket should be fitted with the punch mark facing down and the complete unit positioned so that the punch mark on the handlebars aligns with the joint between the securing bracket and the master cylinder body. Tighten the bracket retaining bolts to a torque of 0.8 – 1.2 kgf m (6 – 9 lbf ft), upper bolt first. On completion of reassembly, refill the reservoir with new hydraulic fluid of the correct specification and bleed the system.

7   The component parts of the master cylinder assembly and the caliper assemblies may wear or deteriorate in function over a long period of use. It is however, generally difficult to foresee how long each component will work with proper efficiency. From a safety point of view it is best to change all the expendable parts every two years on a machine that has covered a normal mileage.

9   Bleeding the front brake hydraulic system

1   If the hydraulic system has to be drained and refilled, if the front brake lever travel becomes excessive or if the lever operates with a soft or spongy feeling, the brakes must be bled to expel air from the system. The procedure for bleeding the hydraulic brake is best carried out by two persons and is as follows.

2   Check the fluid level in the reservoir and if low, top up with new fluid of the correct specification. The reservoir must be kept at least half full of fluid during the bleeding procedure.

3   Ensure the reservoir cap is correctly fitted to prevent a spout of fluid or the entry of dust into the system. Place a clean glass jar below one caliper bleed screw and attach a clear plastic pipe from the caliper bleed screw to the container. Place some clean hydraulic fluid in the jar so that the pipe is always immersed below the surface of the fluid.

4   Operate the brake lever to build up pressure in the system. Once operation of the lever becomes difficult, due to the pressure in the system, unscrew the bleed screw one half turn and squeeze the brake lever. Close the bleed screw just before the distance between the rear face of the end of the lever and the forward face of the throttle twistgrip reaches 15 mm (0.60 in). Do not pull the lever right back to the twistgrip because this will cause piston overtravel resulting in fluid seepage; and do not release the lever until the bleed screw is fully closed otherwise air will be drawn into the system. Repeat the operation until no more air bubbles come from the plastic tube.

5   After the bleeding operation has been completed, check tighten the bleed screw and refit its dust cap. Check the fluid level in the reservoir and repeat the air bleeding procedure with the second caliper unit.

6   Do not use the brake fluid drained from the system, since it will contain minute air bubbles. Never use any fluid other than that recommended. Oil must not be used in any circumstances.

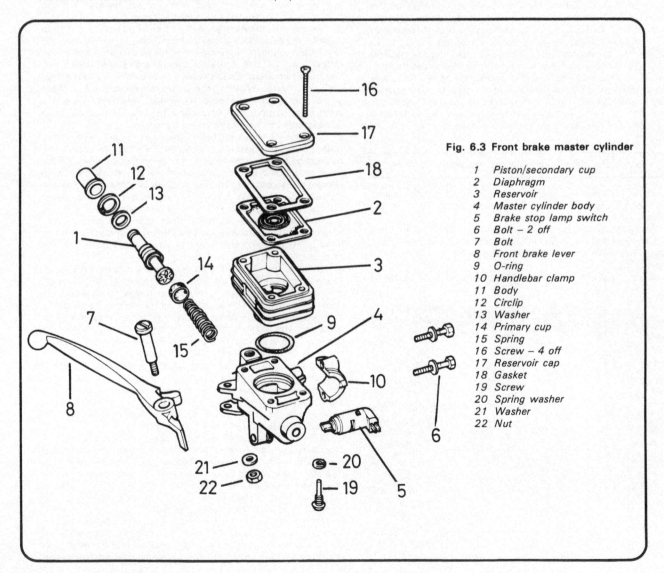

Fig. 6.3 Front brake master cylinder

1   Piston/secondary cup
2   Diaphragm
3   Reservoir
4   Master cylinder body
5   Brake stop lamp switch
6   Bolt – 2 off
7   Bolt
8   Front brake lever
9   O-ring
10 Handlebar clamp
11 Body
12 Circlip
13 Washer
14 Primary cup
15 Spring
16 Screw – 4 off
17 Reservoir cap
18 Gasket
19 Screw
20 Spring washer
21 Washer
22 Nut

10 Rear wheel: examination

1   Place the machine on the centre stand so that the rear wheel is raised clear of the ground. Check for rim alignment damage to the rim and loose or broken spokes by following the procedure relating to the front wheel, as described in Section 2 of this Chapter.

11 Rear wheel: removal and refitting

1   With the machine placed on its centre stand, remove the rear suspension unit lower mounting nuts. Position a block below the wheel in order to prevent the wheel and swinging arm dropping and the wheel spindle fouling the exhaust system.
2   Remove the split-pin from the wheel spindle retaining nut and unscrew and remove the nut. Unscrew and remove the pinch bolt from the left-hand swinging arm end.
3   Before removal of the wheel spindle is attempted, it should be noted that it is possible to bend the left-hand fork leg of the swinging arm during removal of the spindle if the leg is not properly supported against the force required to dislodge the spindle. With the aid of an assistant, use a wooden block to apply inward pressure to the end of the left-hand fork leg of the swinging arm whilst drifting the wheel spindle out from the right.
4   With the spindle removed from the machine and the rear suspension unit lower mounting bolt and stud still supporting the wheel and swinging arm, raise the brake caliper assembly clear of the disc. It is worth fitting a wooden wedge between the brake pads to prevent their expulsion should the brake pedal be operated whilst the wheel is removed. Operating from the left-hand side of the machine, pull the wheel sideways to dislodge the final drive flange from the final drive housing. The wheel can now be manoeuvred clear of the machine by tilting it so as to clear the exhaust silencer and mudguard and then pulling it rearwards.

5   On Interstate models, because of the limited accessibility to the rear wheel imposed by the pannier assembly, it is necessary to employ a different method of rear wheel removal. With the caliper assembly raised clear of the disc and wedged as described above, unscrew and remove the three final drive housing to torque tube retaining nuts and pull the wheel, with the housing attached, rearwards. Detach the final drive housing from the wheel, taking care not to tilt it too much from the vertical otherwise oil will be lost from the unit via the breather. With an assistant tilting the machine to the right, the wheel may be manoeuvred out of position from the left.
6   Refit the wheel by reversing the removal procedure, noting the following points. Before refitting the final drive flange into the final drive housing, lubricate the splines of both components with a multi-purpose lithium based grease. If the final drive flange has been detached from the wheel, the damper pins should also be lubricated with the same grease before re-insertion into the flexible bushes. On Interstate models, lubricate the final drive shaft splines before reconnecting the final drive housing to the torque tube, again with the same grease. Ensure that splines of the final drive housing are correctly aligned with the splines of both the final drive flange and final drive shaft.
7   Carefully lower the brake caliper assembly over the disc so as to avoid damage to the pads and insert the wheel spindle through the swinging arm fork ends, caliper support, wheel hub and final drive housing. Do not omit to fit the spacer collar which locates in the left-hand side of the wheel hub. Loosely fit the spindle retaining nut.
8   Tighten the final drive housing to torque tube retaining nuts evenly to a torque of 3.5 – 4.5 kgf m (25 – 33 lbf ft). Tighten the wheel spindle retaining nut to a torque of 8.0 – 10.0 kgf m (58 – 72 lbf ft) whilst preventing the spindle from turning by inserting a tommy bar through the hole in the end of the spindle. Fit a new split pin to lock the nut in position. Fit and tighten the pinch bolt to a torque of 2.4 – 2.9 kgf m (17 – 21 lbf ft). Fit and tighten the suspension unit lower mounting nuts to a torque of 3.0 – 4.0 kgf m (22 – 29 lbf ft).

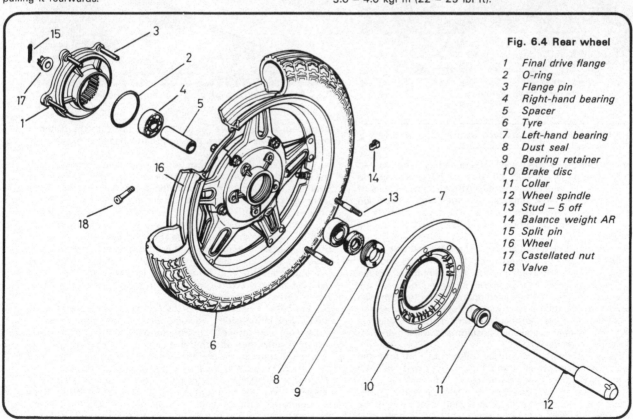

Fig. 6.4 Rear wheel

1   Final drive flange
2   O-ring
3   Flange pin
4   Right-hand bearing
5   Spacer
6   Tyre
7   Left-hand bearing
8   Dust seal
9   Bearing retainer
10   Brake disc
11   Collar
12   Wheel spindle
13   Stud – 5 off
14   Balance weight AR
15   Split pin
16   Wheel
17   Castellated nut
18   Valve

9.3 Use brake bleeder kit, or tube and glass jar, for bleeding operation

11.2 Remove the pinch bolt from the swinging arm end

11.7 Do not omit to refit the spacer collar

12.2 Fabricate a tool with which to remove the bearing retainer

12 Rear wheel bearings: removal, examination and fitting

1　Remove the brake disc, which is retained by five nuts that screw on to studs in the hub. Lift out the final drive splined flange, the pins of which fit into cush drive rubber bushes. If the flange has not been removed for some time, corrosion between the pins and bush sleeves will cause dificulty in removal. A two-legged sprocket puller can be used to aid removal.

2　Remove the bearing retainer from the left-hand side of the wheel hub by using a peg spanner. If the Honda tool No 07710-0010100 is not available, this tool may be fabricated from a length of steel bar and two nuts and bolts of the appropriate diameter (see the accompanying photograph). Do not attempt to drift the retainer loose by using a hammer and punch or similar tools as this will only result in damage occurring. The retainer is staked in position and will require firm even pressure to release it.

3　The left-hand bearing should be drifted out first, from the right-hand side of the wheel. Use a long drift against the inner face of the inner race. It may be necessary to knock the collar to one side so that purchase can be made against the race. Work round in a circle to keep the bearing square in the housing. The dust seal will be pushed out as the bearing is

displaced. After removal of the bearing, take out the distance collar and then drift out the right-hand bearing in a similar manner.

4　Remove all the old grease from the hub and bearings, wash the bearings in petrol, and dry them thoroughly. Check the bearings for roughness by spinning them whilst holding the inner track with one hand and rotating the outer track with the other. If there is the slightest sign of roughness renew them.

5　Before driving the bearings back into the hub, pack the hub with new grease and also grease the bearings. Using a tubular drift of the same diameter as the outer race of each bearing, drive the bearings back into the wheel hub. Ensure that each bearing enters the hub squarely and is fitted with the closed side facing outwards. Do not omit to refit the distance collar before fitting the second bearing.

6　Fit the new dust seal over the left-hand bearing and screw the retainer into position, having obtained a new replacement if there were signs of wear or damage to the threads. Tighten the retainer firmly, then secure it in this position by staking at the junction of the retainer and the wheel hub. A locking fluid may be used to secure the retainer in preference to staking.

7　Refit the brake disc and tighten the nuts evenly and in a diagonal sequence to a torque of 2.7 – 3.3 kgf m (20 – 24 lbf ft). Before refitting the final drive flange, lubricate the damper pins with a multi-purpose lithium based grease.

12.3 Use a long drift to remove the wheel bearings

12.5a Do not omit to refit the distance collar ...

12.5b ... before inserting the second bearing into the wheel hub ...

12.5c ... ensuring that the bearing is driven in squarely

12.6 Stake the bearing retainer to lock it in position

12.7a Tighten the brake disc securing nuts evenly and in a diagonal sequence

12.7b Lubricate the final drive flange damper pins

around the friction surface to denote the extent of allowable wear. These are visible through the small inspection window on the upper face of the caliper. To gain access to the pads for renewal it will be necessary to remove the caliper, leaving the caliper bracket assembly and pads in place. It is not necessary to disturb the hydraulic system.

4   Unscrew the two bolts which retain the caliper to the caliper bracket, and lift the caliper clear. Support the caliper to avoid placing undue strain on the hose or unions. The pads and their anti-squeal shim can now be removed in the same manner as the front caliper components.

5   When fitting new pads it should be noted that the arrow on the shim must face in the direction of normal wheel travel (towards the front of the machine). It will be necessary to depress the piston slightly when fitting the caliper over the new pads. This will allow sufficient clearance.

6   Note that the reservoir level should be checked as the fluid displaced from the caliper may take it above the normal maximum level. Check that the small bellows-type dust seals are in sound condition, because road dirt and water will cause rapid corrosion if it finds its way in. This applies equally to the main piston boot. Tighten the caliper retaining bolts to a torque of 1.5 – 2.0 kgf m (11 – 14 lbf ft).

13 Rear disc brake assembly: examination and brake pad renewal

1   Overhaul and maintenance of the rear brake assembly follows a similar procedure to that described earlier for the front brake components. It follows that similar precautions must be taken to avoid the ingress of dirt, moisture or air into the system. The master cylinder is housed immediately behind the rear brake pedal and is connected by a short low-pressure hydraulic hose to the remote reservoir behind the right-hand side. A high-pressure hydraulic hose runs from the master cylinder to the rear brake caliper.

2   Check the unions and hose connections for signs of leakage, noting that prompt action must be taken if any such signs are noted. Bear in mind that although the hose between the reservoir and master cylinder is not under great pressure, any leakage could allow fluid spillage or even cause air to be admitted into the system.

3   Like the front brake pads, the rear pads carry a red line

14 Rear brake caliper: examination and overhaul

1   The procedure for overhauling the rear brake caliper is essentially the same as that described for the front brake in Section 6 of this Chapter. It should be noted that, in view of its location, the rear caliper is likely to be more seriously affected by the accumulation of road dirt than the front units. Appropriate steps should be taken to prevent the ingress of dirt into the caliper during removal.

2   Slacken the two retaining bolts and lift the caliper clear of the caliper bracket, but leave the hydraulic hose attached at this stage. Remove the bleed valve dust cap, then slacken the bleed valve, holding the caliper inverted over a drain tray. Operate the brake pedal repeatedly to expel the hydraulic fluid. Tighten the bleed valve and disconnect the hydraulic hose. Further dismantling of the caliper follows the dismantling sequence given in Section 6. Note that any work on the hydraulic system must be undertaken under ultra-clean conditions. Particles of dirt will score the working parts and cause early failure.

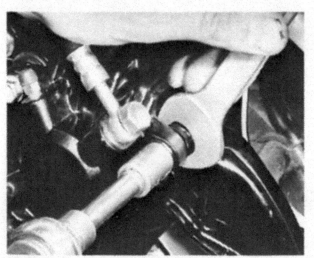

13.4 Release the brake caliper from its support bracket

13.6 Check the fluid level in the reservoir

3 The caliper bracket supports the caliper above the rear disc, allowing a controlled sliding movement to equalise pressure between the pads when the brake is operated. If it is desired to remove the bracket from the machine, remove the torque arm bolt and displace the wheel spindle to allow the bracket to be pulled clear.

4 Remove the caliper sliding pins and check for wear or corrosion. It is imperative that these move smoothly in the bracket if full braking effort is to be expected. Note that wear will allow the caliper to chatter during braking, necessitating renewal of the pins and/or bracket. It follows that the small bellows seals must function effectively if rapid wear is to be avoided. It is worth renewing these components during overhaul, as a precautionary measure. When reassembling the bracket, lubricate the pins with a medium weight high-temperature silicone grease. Always fit a new split-pin to both the torque arm bolt and wheel spindle and spread the legs of the pin to lock if in position. Note the torque figures given in Specifications for the wheel spindle nut and pinch bolt and the torque arm securing nut.

14.2 Lift the brake caliper clear of its support bracket

14.4a Take care not to displace the bush from the caliper support arm torque arm mounting

14.4b Always fit a new split-pin to the torque arm bolt

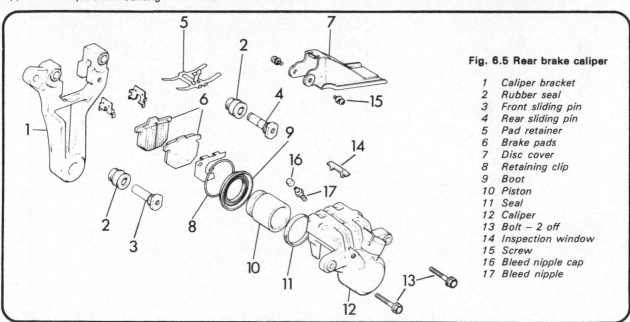

Fig. 6.5 Rear brake caliper

1 Caliper bracket
2 Rubber seal
3 Front sliding pin
4 Rear sliding pin
5 Pad retainer
6 Brake pads
7 Disc cover
8 Retaining clip
9 Boot
10 Piston
11 Seal
12 Caliper
13 Bolt – 2 off
14 Inspection window
15 Screw
16 Bleed nipple cap
17 Bleed nipple

15 Rear brake disc: examination, removal and refitting

1   The rear brake disc should be examined for wear, noting that the likelihood of scoring is higher in the case of the rear item than for its front wheel counterpart, due to the location. Check for wear and warpage as described in Section 7, noting the service limits given in the Specifications.

2   Disc removal is straightforward after the wheel has been removed from the machine as described in Section 11 of this Chapter. This disc is retained on the wheel hub by five nuts which screw in the ends of studs set into the hub casting. These nuts should be undone evenly and in a diagonal sequence and fitted and tightened in the same manner, torque loading them to 2.7 – 3.3 kgf m (20 – 24 lbf ft).

16 Rear brake master cylinder: removal, examination, renovation and refitting

1   Remove the right-hand side panel to gain access to the hydraulic fluid reservoir. Slacken the hose clip which retains the low-pressure hydraulic feed pipe to the master cylinder. Cover the surrounding area with rags to protect the paintwork, then pull off the pipe and allow the reservoir to drain into a suitable receptacle. Discard the fluid, which should not be re-used.

2   Release the master cylinder pushrod from the brake pedal by withdrawing the split-pin and displacing the clevis pin. The master cylinder can now be removed after unscrewing the two retaining bolts which secure it to the frame mounting. Discon-

nect the union between the master cylinder and brake hose and mask the hose end to prevent the ingress of moisture and dirt into the brake system.

3   Pull back the rubber boot to expose the circlip which retains the pushrod assembly. Release the circlip and withdraw the pushrod. The master cylinder piston primary cup and spring can now be removed, noting that it may be necessary to displace them using compressed air or a footpump on the fluid outlet. Take care to avoid damage or injury through fluid splashes during this operation. Wrap the assembly in rag and look away whilst operating the air line or pump. For further details of the overhaul procedure, refer back to Section 8.

4   Note that, if required, the fluid reservoir can be removed from its frame mounting simply by unscrewing the single 6 mm securing bolt. When refitting the reservoir and master cylinder assembly, note the torque figures given in the Specifications of this Chapter. Check also the brake pedal height in accordance with the information given under the 'Brake system inspection' heading in the Routine Maintenance Chapter.

17 Bleeding the rear brake hydraulic system

1   As with the front brake, it will be necessary to bleed the rear hydraulic system whenever there has been a likelihood of air entering it. This will obviously apply if the system has been overhauled or if the caliper or master cylinder have been removed or dismantled. Follow the sequence described in Section 9, operating the rear brake pedal in place of the handlebar lever. Remember to keep the fluid level within limits during the bleeding operation.

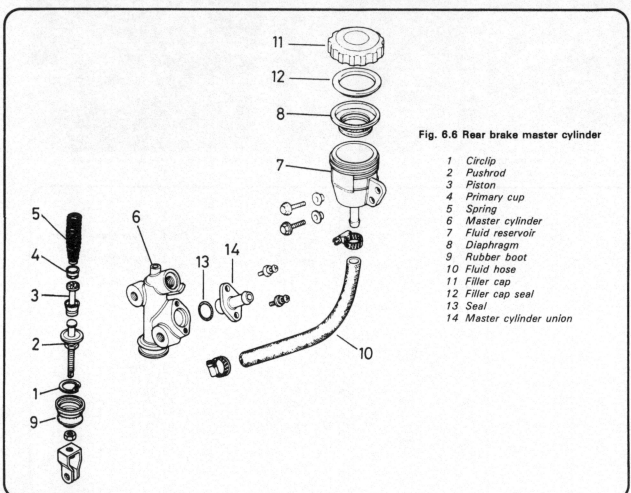

Fig. 6.6 Rear brake master cylinder

1   Circlip
2   Pushrod
3   Piston
4   Primary cup
5   Spring
6   Master cylinder
7   Fluid reservoir
8   Diaphragm
9   Rubber boot
10 Fluid hose
11 Filler cap
12 Filler cap seal
13 Seal
14 Master cylinder union

18 Cush drive assembly: examination

1   A cush drive assembly is incorporated in the rear wheel to absorb any shocks transmitted from the final drive gear to the wheel. The system comprises five flexible bushes inserted in the rear hub which fit the pins of the final drive flange, which is connected to the drive gear by a splined boss.

2   After considerable service the rubber flexible bushes will wear and become compacted, giving rise to excessive backlash between the drive shaft and the wheel. The bushes will then require renewal if original transmission smoothness is to be maintained. The drive flange can be pulled from the wheel hub after the rear wheel has been removed. It is probable the corrosion between the metal bush inserts and the pins will have rendered the flange immovable. A two or three legged sprocket puller can be utilised to aid extraction of the flange.

3   The bushes are not available as separate replacement parts; if they are worn, therefore, the complete rear wheel must be renewed. The only alternative is to find a good Honda dealer who is prepared to compare the bushes with those in his stock; many Honda machines employ a similar cush-drive arrangement, and one or two may use bushes of the same size and strength that are available separately. Bush removal is very difficult and can be achieved only by the use of the correct internally expanding extractor tool; it is recommended that the wheel be taken to a Honda dealer for the work to be carried out. New bushes, if available, can be drifted into the hub after lightly lubricating the inner and outer sleeves with a graphite grease.

19 Final drive system: examination and renovation

1   The final drive housing and drive shaft can be removed from the machine by following the procedure given for swinging arm removal in Chapter 5, Section 9. The only time when the swinging arm does not require removal to extract the final drive shaft is when the engine unit is out of the frame.

2   It is strongly advised that the final drive housing incorporating the final drive crown and pinion gears be returned to a Honda Service Agent if and when servicing or overhaul is required. Dismantling the gear unit requires the use of special tools which are not generally obtainable by the public. Additionally, reassembly of the unit requires working to very close tolerances and the preload adjustment of bearings and gears.

3   Check the drive shaft after removal for slop in the splined joints and for wear in the universal joint. Renew the matching components, where necessary.

18.1a Remove the final drive flange from the rear wheel hub ...

18.1b ... to allow inspection of the flexible bushes

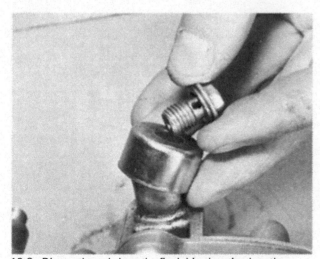

19.0a Dismantle and clean the final drive housing breather

19.0b Check that the vent at the bottom of the final drive housing is kept clear

20 Tyres: removal and refitting

1   It is strongly recommended that should a repair to a tubeless tyre be necessary, the wheel is removed from the machine and taken to a tyre fitting specialist who is willing to do the job or taken to an official Honda dealer. This is because the force required to break the seal between the wheel rim and tyre bead is considerable and considered to be beyond the capabilities of an individual working with normal tyre removing tools. Any abortive attempt to break the rim to bead seal may also cause damage to the wheel rim, resulting in an expensive wheel replacement. If, however, a suitable bead releasing tool is available, and experience has already been gained in its use, tyre removal and refitting can be accomplished as follows.

2   To remove the tyre from either wheel, first detach the wheel from the machine by following the procedure in Sections 3 or 11 depending on whether the front or the rear wheel is involved. Deflate the tyre by removing the valve insert and when it is fully deflated, push the bead of the tyre away from the wheel rim on both sides so that the bead enters the centre well of the rim. As noted, this operation will almost certainly require the use of a bead releasing tool.

3   Insert a tyre lever close to the valve and lever the edge of the tyre over the outside of the wheel rim. Very little force should be necessary; if resistance is encountered it is probably due to the fact that the tyre beads have not entered the well of the wheel rim all the way round the tyre. Should the initial problem persist, lubrication of the tyre bead and the inside edge and lip of the rim will facilitate removal. Use a recommended lubricant, a dilute solution of washing-up liquid or french chalk. Lubrication is usually recommended as an aid to tyre fitting but its use is equally desirable during removal. The risk of lever damage to wheel rims can be minimised by the use of proprietary plastic rim protectors placed over the rim flange at the point where the tyre levers are inserted. Suitable rim projectors may be fabricated very easily from short lengths (4-6 inches) of thick-walled nylon petrol pipe which have been split down one side using a sharp knife. The use of rim protectors should be adopted whenever levers are used and, therefore, when the risk of damage is likely.

4   Once the tyre has been edged over the wheel rim, it is easy to work around the wheel rim so that the tyre is completely free on one side.

5   Working from the other side of the wheel, ease the other edge of the tyre over the outside of the wheel rim, which is furthest away. Continue to work around the rim until the tyre is freed completely from the rim.

6   Refer to the following Section for details relating to puncture repair and the renewal of tyres. See also the remarks relating to the tyre valves in Section 22.

7   Refitting of the tyre is virually a reversal of removal procedure. If the tyre has a balance mark (usually a spot of coloured paint), as on the tyres fitted as original equipment, this must be positioned alongside the valve. Similarly, any arrow indicating direction of rotation must face the right way.

8   Starting at the point furthest from the valve, push the tyre bead over the edge of the wheel rim until it is located in the central well. Continue to work around the tyre in this fashion until the whole of one side of the tyre is on the rim. It may be necessary to use a tyre lever during the final stages. Here again, the use of a lubricant will aid fitting. It is recommended strongly that when refitting the tyre only a recommended lubricant is used because such lubricants also have sealing properties. Do not be over generous in the application of lubricant or tyre creep may occur.

9   Fitting the upper bead is similar to fitting the lower bead. Start by pushing the bead over the rim and into the well at a point diametrically opposite the tyre valve. Continue working round the tyre, each side of the starting point, ensuring that the bead opposite the working area is always in the well. Apply lubricant as necessary. Avoid using tyre levers unless absolutely

essential, to help reduce damage to the soft wheel rim. The use of the levers should be required only when the final portion of bead is to be pushed over the rim.

10   Lubricate the tyre beads again prior to inflating the tyre, and check that the wheel rim is evenly positioned in relation to the tyre beads. Inflation of the tyre may well prove impossible without the use of a high pressure air hose. The tyre will retain air completely only when the beads are firmly against the rim edges at all points and it may be found when using a foot pump that air escapes at the same rate as it is pumped in. This problem may also be encountered when using an air hose, on new tyres which have been compressed in storage and by virtue of their profile hold the beads away from the rim edges. To overcome this difficulty, a tourniquet may be placed around the circumference of the tyre, over the central area of the tread. The compression of the tread in this area will cause the beads to be pushed outwards in the desired direction. The type of tcurniquet most widely used consists of a length of hose closed at both ends with a suitable clasp fitted to enable both ends to be connected. An ordinary tyre valve is fitted at one end of the tube so that after the hose has been secured around the tyre it may be inflated, giving a constricting effect. Another possible method of seating beads to obtain initial inflation is to press the tyre into the angle between a wall and the floor. With the airline attached to the valve additional pressure is then applied to the tyre by the hand and shin, as shown in the accompanying illustration. The application of pressure at four points around the tyre's circumference whilst simultaneously applying the airhose will often effect an initial seal between the tyre beads and wheel rim, thus allowing inflation to occur.

11   Having successfully accomplished inflation, increase the pressure to 40 psi and check that the tyre is evenly disposed on the wheel rim. This may be judged by checking that the thin positioning line found on each tyre wall is equidistant from the rim around the total circumference of the tyre. If this is not the case, deflate the tyre, apply additional lubrication and reinflate. Minor adjustments to the tyre position may be made by bouncing the wheel on the ground.

12   Always run the tyre at the recommended pressures and never under or over-inflate. The correct pressures for solo use are given in the Specification Section of this Chapter. If a pillion passenger is carried, increase the rear tyre pressure only as recommended.

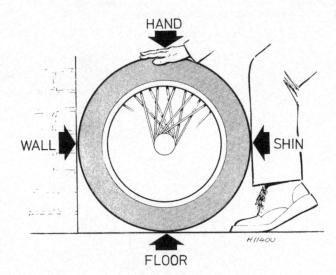

Fig. 6.7 Method of tubeless tyre inflation

Apply pressure at the four marked points to facilitate initial inflation

TYRE CHANGING SEQUENCE - TUBELESS TYRES

Deflate tyre. After releasing beads, push tyre bead into well of rim at point opposite valve. Insert lever next to valve and work bead over edge of rim.

Use two levers to work bead over edge of rim. Note use of rim protectors.

When first bead is clear, remove tyre as shown.

Before installing, ensure that tyre is suitable for wheel. Take note of any sidewall markings such as direction of rotation arrows.

Work first bead over the rim flange.

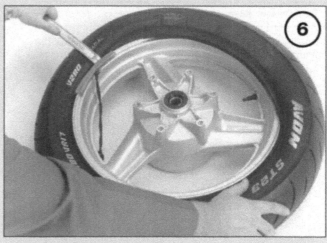

Use a tyre lever to work the second bead over rim flange.

21 Puncture repair and tyre renewal

1   The primary advantage of the tubeless tyre is its ability to accept penetration by sharp objects such as nails etc without loss of air. Even if loss of air is experienced, because there is no inner tube to rupture, in normal conditions a sudden blow-out is avoided.

2   If a puncture of the tyre occurs, the tyre should be removed for inspection for damage before any attempt is made at remedial action. The temporary repair of a punctured tyre by inserting a plug from the outside should not be attempted. Although this type of temporary repair is used widely on cars, the manufacturers strongly recommend that no such repair is carried out on a motorcycle tyre. Not only does the tyre have a thinner carcass, which does not give sufficient support to the plug, but the consequences of a sudden deflation is often sufficiently serious that the risk of such as occurrence should be avoided at all costs.

3   The tyre should be inspected both inside and out for damage to the carcass. Unfortunately the inner lining of the tyre – which takes the place of the inner tube – may easily obscure any damage and some experience is required in making a correct assessment of the tyre condition.

4   There are two main types of tyre repair which are considered safe for adoption in repairing tubeless motorcycle tyres. The first type of repair consists of inserting a mushroom-headed plug into the hole from the **inside** of the tyre. The hole is prepared for insertion of the plug by reaming and the application of an adhesive. The second repair is carried out by buffing the inner lining in the damaged area and applying a cold or vulcanised patch. Because both inspection and repair, if they are to be carried out safely, require experience in this type of work, it is recommended that the tyre be placed in the hands of a repairer with the necessary skills, rather than repaired in the home workshop.

5   In the event of an emergency, the only recommended 'get-you-home' repair is to fit a standard inner tube of the correct size. If this course of action is adopted, care should be taken to ensure that the cause of the puncture has been removed before the inner tube is fitted. It will be found that the valve hole in the rim is considerably larger than the diameter of the inner tube valve stem. To prevent the ingress of road dirt, and to help support the valve, a spacer should be fitted over the valve. A conversion spacer for most Honda models equipped with Comstar wheels is available from Honda dealers.

6   In the event of the unavailability of tubeless tyres, ordinary tubed tyres fitted with inner tubes of the correct size may be fitted. Refer to the manufacturer or a tyre fitting specialist to ensure that only a tyre and tube of equivalent type and suitability is fitted, and also to advise on the fitting of a valve nut/spacer to the rim hole.

22 Tyre valves: description and renewal

1   It will be appreciated from the preceding Sections that the adoption of tubeless tyres has made it necessary to modify the valve arrangement, as there is no longer an inner tube which can carry the valve core. The problem has been overcome by using a moulded rubber valve body which locates in the wheel rim hole. The valve body is pear-shaped, and has a groove around its widest point which engages with the rim forming an airtight seal.

2   The valve is fitted from the rim well, and it follows that it can only be renewed whilst the tyre itself is removed from the wheel. Once the valve has been fitted, it is almost impossible to remove it without damage, and so the simplest method is to cut it as close as possible to the rim well. The two halves of the old valve can then be removed.

3   The new valve is fitted by inserting the threaded end of the valve body through the rim hole, and pulling it through until the groove engages in the rim. In practice, a considerable amount of pressure is required to pull the valve into position, and most tyre fitters have a special tool which screws onto the valve end to enable purchase to be obtained. It is advantageous to apply a little tyre bead lubricant to the valve to ease its insertion. Check that the valve is seated evenly and securely.

4   The incidence of valve body failure is relatively small, and leakage only occurs when the rubber valve case ages and begins to perish. As a precautionary measure, it is advisable to fit a new valve when a new tyre is fitted. This will preclude any risk of the valve failing in service. When purchasing a new valve, it should be noted that a number of different types are available. The correct type for use in the Comstar wheel is a Schrader 413, Bridgeport 183M or equivalent.

5   The valve core is of the same type as that used with tubed tyres, and screws into the valve body. The core can be removed with a small slotted tool which is normally incorporated in plunger type pressure gauges. Some valve dust caps incorporate a projection for removing valve cores. Although tubeless tyre valves seldom give trouble, it is possible for a leak to develop if a small particle of grit lodges on the sealing face. Occasionally, an elusive slow puncture can be traced to a leaking valve core, and this should be checked before a genuine puncture is suspected.

6   The valve dust caps are a significant part of the tyre valve assembly. Not only do they prevent the ingress of road dirt into the valve, but also act as a secondary seal which will reduce the risk of sudden deflation if a valve core should fail.

23 Wheel balancing

1   It is customary on all high performance machines to balance the wheels complete with tyre and tube. The out of balance forces which exist are eliminated and the handling of the machine is improved in consequence. A wheel which is badly out of balance produces through the steering a most unpleasant hammering effect at high speeds.

2   Some tyres have a balance mark on the sidewall, usually in the form of a coloured dot. This mark must be in line with the tyre valve, when the tyre is fitted to the inner tube. Even then, the wheel may require the addition of balance weights, to offset the weight of the tyre valve itself.

3   If the wheel is raised clear of the ground and is spun, it will probably come to rest with the tyre valve or the heaviest part downward and will always come to rest in the same position. Balance weights must be added to a point diametrically opposite this heavy spot until the wheel will come to rest in ANY position after it is spun.

4   The Comstar wheels fitted to the GL1100 models have two weights of wheel balancing weight available. These balance weights are available in 20 gm (0.04 lb) and 30 gm (0.07 lb) sizes, which clip to the rim of the wheel.

5   Note that because of the drag of the final drive components the rear tyre must be balanced with the wheel off the machine, supported on a suitable spindle.

24 Fault diagnosis: wheels, brakes and tyres

Symptom	Cause	Remedy
Handlebars oscillate at low speed	Buckle or flat in wheel rim most probably front wheel	Check rim for damage by spinning wheel Renew wheel if not true
	Tyre pressure incorrect	Check, and if necessary adjust
	Tyre not straight on rim	Check tyre fitting. If necessary, deflate tyre and reposition
	Worn wheel or steering bearings	Check and renew or adjust
Machine tends to weave	Tyre pressure incorrect	Check, and if necessary, adjust. If sudden, check for puncture
	Suspension worn or damaged	Check action of front forks and rear suspension units. Check swinging arm for wear
Machine lacks power and accelerates poorly	Front brake binding	Hot disc or caliper indicates binding Overhaul caliper and master cylinder, fit new pads if required, check disc for scoring or warpage
	Rear brake binding	As above
Brakes grab or judder when applied gently	Brake pads badly worn or scored	Renew pads. Check disc and caliper
	Wrong type of pads fitted	Renew
	Warped disc	Renew
Brakes squeal	Glazed pads. Pads worn to backing metal	Sand paper surface to remove glaze, then use brake gently for about 100 miles to permit bedding in. If worn to backing check that disc is not damaged and renew as necessary
	Caliper and pads polluted with brake dust or foreign matter	Dismantle and clean. Overhaul caliper where necessary
Excessive brake lever travel	Air in system	Find cause of air's presence. If due to leak, rectify, then bleed brake
	Very badly worn pads	Renew, and overhaul system where required
	Badly polluted caliper	Dismantle and clean
Brake lever feels springy	Air in system	See above
	Pads glazed	See above
	Caliper jamming	Dismantle and overhaul
Brake pull-off sluggish	Sticking pistons in brake caliper	Overhaul caliper unit
Tyres wear more rapidly in middle of tread	Over-inflation	Check pressure and run at recommended pressures
Rear wheel does not rotate freely	Wheel bearings damaged	Renew
	Rear brake binding	See above
	Wheel spindle bent	Renew
	Swinging arm fork bent	Renew
	Ring and pinion gear bearings of final drive housing damaged	Return to Honda Service Agent
	Excessive preload on final gear assembly	Return to Honda Service Agent
Excessive noise from final drive housing	Oil level low	Replenish with oil of correct grade
	Excessive backlash between pinion and ring gear	Return to Honda Service Agent
	Driven flange and wheel hub damaged	Inspect and renew
	Drive spline damaged	Inspect and renew
	Worn pinion and ring gears	Return to Honda Service Agent
	Ring gear shaft and driven flange worn or damaged	Return to Honda Service Agent
Oil leaking from final drive housing	Oil level too high	Drain oil from housing
	Breather clogged	Remove and clean
	Seals damaged	Inspect and renew

Chapter 7 Electrical system

Contents

Specifications

Alternator
Type .... Three phase, series wound
Output .... 12V, 0.3kW at 5000 rpm

Battery
Make .... YUASA
Type .... Y50 - N18L - A
Voltage 12V
Capacity .... 20 Ah
Polarity .... Negative earth

Voltage regulator/rectifier
Type .... Integrated circuit, non-adjustable

Starter motor
Brush length .... 12 - 13 mm (0.4724 - 0.5118 in)
Service limit .... 5.5 mm (0.2165 in)

Fan motor
Current/speed:
  Without blades .... 1.4A max at 2800 rpm
  With blades .... 4.5 ± 0.5A at 2500 ± 300 rpm

Fuses
Main (battery) 30A
Horn/stop lamp/indicators 15A
Headlamp 10A
Tail lamp 5A
Parking lamp 5A
Instrument warning lights 5A
Suspension air pressure warning light 5A

Bulbs

	USA	UK
Headlamp	H4 (Phillips 12342/99 or equivalent) 12V, 8/23W	H4 (Phillips 12342/99 or equivalent) 12V, 5/21W
Tail/stop lamp	12V, 8/23W	12V, 5/21W
Direction indicator lamps:		
Interstate:		
Front	12V, 8/23W	12V, 5/21W
Rear	12V, 23W	12V, 21W
Standard:		
Front	12V, 8/23W	12V, 21W
Rear	12V, 23W	12V, 21W
Parking lamp	Not fitted	12V, 4W
Instrument illumination lights	12V, 3.4W	12V, 3.4W
Neutral indicator light	12V, 3.4W	12V, 3.4W
Direction indicator warning light	12V, 3.4W	12V, 3.4W
High beam indicator light	12V, 3.4W	12V, 3.4W
Oil pressure warning light	12V, 3.4W	12V, 3.4W
Suspension air pressure warning light	12V, 3.4W	12V, 3.4W

1 General description

The Honda GL 1100 is fitted with a 12V negative earth electrical system. Power is provided by a three-phase fixed-coil alternator, mounted on a cush-drive double gear shaft, which is driven from the crankshaft. The ac (alternating current) output from the alternator is converted to dc (direct current) by a silicon rectifier and is controlled by a solid state voltage regulator. The two components are combined into a single unit. The regulated and rectified supply is then fed to the battery and electrical system.

2 Testing the electrical system: general

1 Checking the electrical output and the performance of the various components within the charging system requires the use of test equipment of the multi-meter type and also an ammeter of 0 - 20 ampere range. When carrying out checks, care must be taken to follow the procedures laid down and so prevent inadvertent incorrect connections or short circuits. Irreparable damage to individual components may result if reversal of current or shorting occurs. It is advised that unless some previous experience has been gained in auto-electrical testing the machine be returned to a Honda Service Agent or auto-electrician, who will be qualified to carry out the work and have the necessary test equipment.

2 If the performance of the charging system is suspect the system as a whole should be checked first, followed by testing of the individual components to isolate the fault. The three main components are the alternator, the rectifier and the regulator; the latter two components being combined. Before commencing the tests, ensure that the battery is fully charged, as described in the following Section.

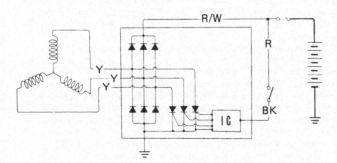

Fig. 7.1 Charging circuit

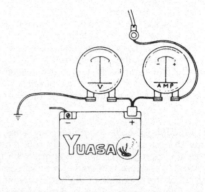

Fig. 7.2 Charging system output test connections

3 Battery: examination and maintenance

1 The battery is housed in a tray located behind the left-hand sidepanel and is retained in position by a bracket which is hinged on the base of the tray and locked in position with a single slotted bolt.

2 The transparent plastic case of the battery permits the upper and lower levels of the electrolyte to be observed without disturbing the battery by removing the side cover. Maintenance is normally limited to keeping the electrolyte level between the prescribed upper and lower limits and making sure that the vent tube is not blocked. The lead plates and their separators are also visible through the transparent case, a further guide to the general condition of the battery. If electrolyte level drops rapidly, suspect over-charging and check the system.

3 Unless acid is spilt, as may occur if the machine falls over, the electrolyte should always be topped up with distilled water to restore the correct level. If acid is spilt onto any part of the machine, it should be neutralised with an alkali such as washing soda or baking powder and washed away with plenty of water, otherwise serious corrosion will occur. Top up with sulphuric acid of the correct specific gravity (1.260 to 1.280) only when spillage has occurred. Check that the vent pipe is well clear of the frame or any of the other cycle parts.

4 It is seldom practicable to repair a cracked battery case because the acid present in the joint will prevent the formation of an effective seal. It is always best to renew a cracked battery, especially in view of the corrosion which will be caused if the acid continues to leak.

5 If the machine is not used for a period of time, it is advisable to remove the battery and give it a 'refresher' charge every six weeks or so from a battery charger. The battery will require recharging when the specific gravity falls below 1.260 (at 20°C

- 68°F). The hydrometer reading should be taken at the top of the meniscus with the hydrometer vertical. If the battery is left discharged for too long, the plates will sulphate. This is a grey deposit which will appear on the surface of the plates, and will inhibit recharging. If there is sediment on the bottom of the battery case, which touches the plates, the battery needs to be renewed. Limit the charging rate to 2A. If charging from an external source with the battery on the machine, disconnect the leads, or the rectifier will be damaged.

6　Note that when moving or charging the battery, it is essential that the following basic safety precautions are taken:

(a) Always remove the vent caps when recharging a battery, otherwise the gas created within the battery when charging takes place will explode and burst the case with disastrous consequences.

(b) Never expose a battery on charge to naked flames or sparks. The gas given off by the battery is highly explosive.

(c) If charging the battery in an enclosed area, ensure that the area is well ventilated.

(d) Always take great care to protect yourself against accidental spillage of the sulphuric acid contained within the battery. Eyeshields should be worn at all times. If the eyes become contaminated with acid they must be flushed with fresh water immediately and examined by a doctor as soon as possible. Similar attention should be given to a spillage of acid on the skin.

Note also that although, should an emergency arise, it is possible to charge the battery at a more rapid rate than that stated in the preceding paragraph, this will shorten the life of the battery and should therefore be avoided if at all possible.

7　Occasionally, check the condition of the battery terminals to ensure that corrosion is not taking place, and that the electrical connections are tight. If corrosion has occurred, it should be cleaned away by scraping with a knife and then using emery cloth to remove the final traces. Remake the electrical connections whilst the joint is still clean, then smear the assembly with petroleum jelly (NOT grease) to prevent recurrence of the corrosion. Badly corroded connections can have a high electrical resistance and may give the impression of complete battery failure.

4 Fuses: location

1　The fuses are contained within a plastic box, mounted on the frame top tubes beneath the dummy fuel tank. They are fitted to give the electrical components protection from sudden overload as occurs in a short circuit. An additional fuse is contained in a box clipped to the rear of the battery tray. This is the main fuse for the battery/starter circuit.

2　If a fuse blows it should not be renewed until the cause of the short is found. This will involve checking the electrical circuit to correct the fault. If this rule is not observed, the fuse will almost certainly blow again.

3　Always carry at least one spare fuse of each type. The main fuse is of the 'spade' type. The fuses carried in the fuse box are of the cylindrical type. Never use a fuse of a higher rating than specified or its protective function will be lost.

4　When a fuse blows whilst the machine is running and no spare fuse is available a 'get you home' remedy is to remove the blown fuse and wrap it in silver paper, this will restore the electrical continuity by bridging the broken wire within the fuse. This expedient should not be used if there is evidence of a short circuit or major electrical fault, otherwise more serious damage will be caused. Replace the 'doctored' fuse at the earliest possible opportunity to restore full circuit protection.

4.1a The fuses are contained within a frame mounted box

4.1b The main fuse is of the ribbon type

5 Alternator: checking output and continuity

1　If the output of the charging system is suspect, it may be checked as follows by using an ammeter and a multimeter connected into the circuit. The ammeter should have a scale of 0 - 5 amps and the multimeter set to the 0 - 20 volts dc scale (or higher). Connect the multimeter across the positive (+) and negative (–) terminals of the battery. Disconnect the positive lead of the battery and reconnect it through the ammeter.

2　Start the engine and allow it to run for five minutes so that it reaches normal running temperature. Increase the engine speed to 3000 rpm and note the readings which should be 3 amps and 14.5 volts. This test should be made with the dip switch on 'main' beam, the fan motor off and with the battery fully charged.

3　If the output of the alternator is erratic or noticeably below the specified amount, check the continuity of the stator coils as follows, by using a multimeter set to the resistance function. Disconnect the alternator main lead at the block connector. Check the continuity between all three of the yellow wires. If there is lack of continuity, there is an open circuit in the coils. Check the continuity between each yellow wire and an earth point. If continuity exists, there is a short in the coils. In either case the defective alternator coil must be renewed.

4　If the alternator wiring is found to be correct but output is incorrect, suspect the regulator/rectifier unit. Test this component as described in the following Section.

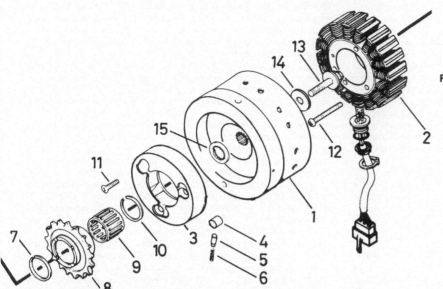

Fig. 7.3 Alternator and starter clutch

1   Rotor
2   Stator
3   Starter clutch
4   Roller – 3 off
5   Cap – 3 off
6   Spring – 3 off
7   Circlip
8   Starter sprocket
9   Needle roller
10 Circlip
11 Screw – 3 off
12 Screw – 3 off
13 Bolt
14 Washer
15 Splined washer

5.3 The alternator stator coils are retained by three screws

6 Regulator/rectifier unit: testing

1   If performance of the charging system is suspect, but the alternator is found to be in good condition, it is probable that one side of the combined regulator/rectifier unit is malfunctioning. Exactly which side is malfunctioning is of academic interest only because the sealed unit cannot be repaired, but must be renewed.

2   Voltage regulator performance may be checked using a voltmeter connected across the battery. Connect the positive voltmeter lead to the positive (+) battery terminal and the negative voltmeter lead to the negative (–) battery terminal. Start the engine and increase the speed until a 14 volt output is registered. Increase the engine speed to approximately 5000 rpm. If the regulator is functioning correctly, the voltage will not rise to above 15 volts. If this voltage is exceeded, the unit is malfunctioning.

3   A further test of the unit's condition may be made by measuring the resistance of the rectifier circuit. To check these values accurately a meter or meters capable of reading in ohms and kilo ohms will be required. Failing this, most multimeters will be able to give a reliable indication despite reading solely on the latter scale. It is generally sufficient to distinguish between

a very high resistance and a very low resistance when checking diode condition.

4   Before the test is started, it is helpful to understand what the rectifier is required to do. We have established that it converts ac to dc for charging purposes, and it does this by a matrix of diodes. Diodes, respresented in diagrammatical form by a triangular symbol with a bar across one end, can be imagined to act as one way valves, passing current in one direction only. Therefore, in a diagram, the current should flow in the direction indicated by the point of the diode symbol, but must not flow back past the bar. In the event of rectifier failure, one or more of the diodes will have failed to operate correctly and thus will either pass current in both directions or will set up a high resistance in both directions. The purpose of the test is thus to check that the diodes are functioning correctly.

5   Trace the leads from the finned regulator/rectifier unit, located beneath the left-hand side of the dummy fuel tank, to the 8-pin block connector mounted just beneath the unit. Separate the two halves of the connector and carry out the following tests.

6.5 The regulator/rectifier unit and block connector

6   Using an ohmmeter or a multimeter set to the resistance function, check the resistance between the green lead and each of the three yellow leads in turn. In the normal direction of current flow the diodes tested should offer little resistance, the precise figure being 5 – 40 ohms. If the test probes are reversed, however, a very high resistance should be shown, approximately 6000 ohms (6 k ohms).

7   Repeat the above tests, but this time check the resistance between the red/white lead and each of the yellow leads in turn. Again, a low resistance of 5 – 40 ohms should be shown, with a high value of approximately 6000 ohms with the probes reversed. It follows that if in any of the tests the meter indicates a very high or very low resistance in both directions, the unit is defective and must be renewed.

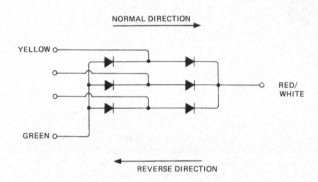

Fig. 7.4 Rectifier circuit resistance check

5   Before the brushes are replaced, make sure the commutator on which they bear is clean. If necessary, the commutator segments may be burnished using Crokus paper. This is a fine abrasive paper produced for this purpose, and can be obtained from auto-electrical specialists. On no account use emery paper as it will damage the commutator. After cleaning, wipe the commutator with a rag soaked in methylated spirits or de-natured alcohol to remove any dust or grease.

6   The insulation between the individual segments of the commutator should lie a minimum distance of 2 mm (0.08 in) below the surface. If the insulation is less than this distance below the surface, it should be undercut very carefully, using a small portion of fine hacksaw blade. It may be considered worthwhile to seek the services of an auto-electrician who will be able to carry out this delicate operation at a favourable price.

7   Closely inspect the individual bars of the commutator for signs of discolouration. Any discolouration between one pair of bars indicates earthed or open armature coils, whereas discolouration on the edge of a bar indicates a high resistance. Using a multimeter set to the resistance testing function, test for continuity between the pairs of bars and between each bar and the armature shaft. Check also for continuity between the cable terminal on the starter motor body and the motor body itself and also between the cable terminal on the motor body and the wire connecting the field coil to the brushes. If the armature coils are found to have an open circuit or are shorted to the armature shaft, or if the field coil does not show continuity or is shorted to the motor body, the starter motor should be returned to an official Honda Service Agent or an auto-electrician for repair or replacement.

8   If the motor is serviceable, fit the brushes in their holders and check that they slide quite freely. Make sure that used brushes are refitted in their original positions because they will have worn to the profile of the commutator. Reassemble and refit the starter motor using a reversal of the removal procedure. Note that before fitting the motor to the engine, the O-ring on the motor body must be inspected for signs of damage or deterioration and renewed if necessary.

7   Starter motor: removal, examination and fitting

1   A push button switch is located on the right-hand section of the handlebars. When depressed, it operates a solenoid which in turn causes the starter motor to operate. This drives a free running clutch via a chain, which in turn rotates an idler upon which the alternator is mounted, and finally the crankshaft.

2   To remove the starter motor, first detach the negative terminal at the battery and the main lead on the starter motor body. Remove the left-hand exhaust pipe and gear change lever. The starter motor is retained by two bolts passing into the crankcase. After removal of the bolts the motor can be eased backwards and away from the machine.

3   The parts of the starter motor most likely to require attention are the brushes. The end cover is retained by the two long screws which pass through the lugs cast on both end pieces. If the screws are withdrawn, the end cover can be lifted away and the brush gear exposed.

4   Lift up the spring clips which bear on the end of each brush and remove the brushes from their holders. The standard length and wear limit of the brushes is given in the Specifications at the beginning of this Chapter. If either brush is worn to a length less than that of the wear limit, renew the brushes as a pair.

7.3 Remove the starter motor end cover to expose the brush gear

7.4 Displace the retaining springs and pull the brushes from their holders

8 Starter solenoid switch: function and testing

1 The starter motor switch is designed to work on the electro-magnetic principle. When the starter motor button is depressed, current from the battery passes through windings in the switch solenoid and generates an electro-magnetic force which causes a set of contact points to close. Immediately the points close, the starter motor is energised and a very heavy current is drawn from the battery.

2 This arrangement is used for two reasons. Firstly, the starter motor current is drawn only when the button is depressed and is cut off again when pressure on the button is released. This ensures minimum drainage on the battery. Secondly, if the battery is in a low state of charge, there will not be sufficient current to cause the solenoid contacts to close. In consequence it is not possible to place an excessive drain on the battery which in some circumstances, can cause the plates to overheat and shed their coatings. If the starter will not operate, first suspect a discharged battery. This can be checked by trying the horn or switching on the lights. If this check shows the battery to be in good shape, suspect the starter switch which should come into action with a pronounced click. It is located under the left-hand sidepanel, directly to the rear of the battery. Before condemning the starter solenoid, carry out the following tests.

3 Disconnect the earth lead from the negative (–) terminal of the battery and move it well clear of the terminal. Remove the rubber cover from the starter solenoid and unplug the 4-pin block connector from the solenoid body. Unscrew and remove the two nuts from the solenoid and disconnect the leads from the threaded terminals. Remove the starter solenoid from its retainer by sliding it out of position.

4 With a multimeter set to the resistance function, test the primary coil of the solenoid for the correct resistance. This is done by connecting the probes for the multimeter to the two connector tabs nearest the two threaded terminals. The resistance reading obtained should be 3.4 ± 10% ohms. Test also for shorting between either of the connector tabs and the solenoid body. If shorting is found or if the resistance reading obtained is incorrect, the solenoid should be replaced with a serviceable item.

5 Energise the solenoid by connecting a 12 volt battery to the two aforementioned primary coil connector tabs. The relay should make a pronounced click as the second of the two leads is connected to the tab. With the solenoid energised, connect the probes of a multimeter, set on the resistance function, to the two threaded terminals of the secondary coil. Test for continuity across the terminals; if no continuity exists, the solenoid must be replaced with a serviceable item.

6 If these tests prove satisfactory, check that the system malfunction is not caused by a defective clutch interlock switch or neutral interlock switch. Unless either or both of these are working, ie neutral has been selected and/or the clutch is disengaged, the starter will not operate. Check the operation of both switches as described later in this Chapter.

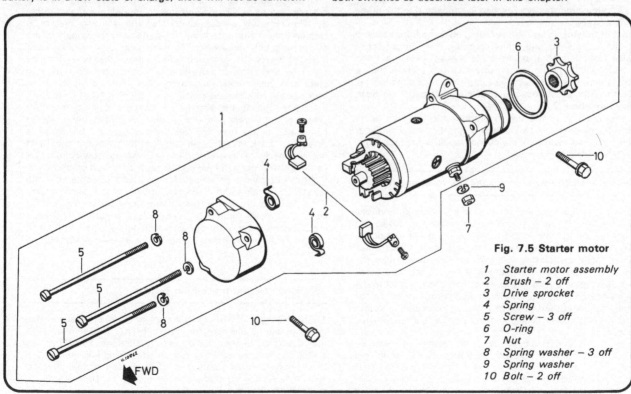

Fig. 7.5 Starter motor

1 Starter motor assembly
2 Brush – 2 off
3 Drive sprocket
4 Spring
5 Screw – 3 off
6 O-ring
7 Nut
8 Spring washer – 3 off
9 Spring washer
10 Bolt – 2 off

9 Ignition switch: removal, testing and fitting

1   The combined ignition and lighting master switch is bolted to the upper yoke, and may be removed after the instrument panel and headlamp shell have been detached. Disconnect the block connector plug from the base of the ignition switch. The switch is held in place by two screws, after the removal of which the switch can be displaced downwards.
2   Repair of a malfunctioning switch is not practicable; renewal being the only solution. The switch may be refitted by reversing the dismantling sequence. Remember that a new switch will also require a new set of keys.

10 Handlebar switches

1   To remove the handlebar switches, first disconnect the battery and also the leads to the switches. Both switches can be detached from the handlebars after separating the halves which are clamped together from underneath by retaining screws.
2   The throttle cables will have to be disconnected from the right-hand switch. This can be accomplished by slackening the adjusters off at both ends to give the required amount of play.
3   Disconnect the cable connection from the clutch lever assembly. Note also the small switch on the clutch lever which isolates the solenoid and so prevents starting of the engine when the machine is in gear and the clutch is not disengaged.
4   Repair of either switch assembly is seldom practicable. If the switches malfunction due to corrosion or dirt on the contacts they can be cleaned most easily with a proprietary electrical contact cleaner which can be sprayed onto the affected areas.

11 Stop lamp switches: adjustment

1   The rear brake lever operated stop lamp switch is located forward of and above the right-hand swinging arm pivot; it has a threaded body giving a range of adjustment. The switch is held in a frame-mounted bracket and is connected to the brake pedal via a spring.
2   If the stop lamp is late in operating, hold the body still and rotate the combined adjusting/mounting nut in an anti-clockwise direction so that the body moves away from the brake pedal shaft. If the switch operates too early or has a tendency to stick on, rotate the nut in a clockwise direction. As a guide, the light should operate after the brake pedal has been depressed by about 2 cm (0.75 in).
3   A stop lamp switch is also incorporated in the front brake system. The mechanical switch is a push fit in the handlebar lever stock. If the switch malfunctions, repair is impracticable. The switch should be renewed.

12 Neutral indicator switch: testing, removal and fitting

1   The neutral indicator switch is fitted in the right-hand wall of the crankcase, forward of the oil strainer access cover, and operates a warning lamp in the instrument console to indicate that neutral has been selected. More importantly, it is interconnected with the starter solenoid and will only allow the engine to be started if the gearbox is in neutral, unless the clutch is disengaged. It can be checked by setting a multimeter on the resistance scale and connecting one probe to the switch terminal and the other to earth. The meter should indicate continuity when neutral is selected and infinite resistance when in any gear.
2   If the switch is found to be unserviceable, it must be renewed. Drain the engine oil to lower the level in the crankcase, and remove the right-hand exhaust pipe so that access to the switch can be gained. Remove the bolt and claw plate that retains the switch and remove the front right-hand

engine mounting nut. In order that the switch clears the engine mounting lug the frame must be prised outwards about 2.0 mm (0.08 in). This can be accomplished by placing a wooden lever between the engine and frame. Pull the switch out until it leaves the crankcase.
3   Refit and reconnect the new switch by reversing the removal procedure.

13 Clutch interlock switch: general

1   A small plunger-type switch is incorporated in the clutch lever, serving to prevent operation of the starter circuit when any gear has been selected, unless the clutch lever is held in. It can be checked by the method described above for the neutral switch. The meter should indicate continuity with the clutch applied and infinite resistance with the clutch released.
2   If defective, the switch must be renewed, as there is no satisfactory means of repair. The switch can be removed after releasing the clutch cable and lever blade.

14 Rear suspension air pressure warning light, light control unit and pressure sensor switch: location and testing

1   The warning light for the rear suspension air pressure is located within the face of the tachometer and should illuminate for 5 seconds immediately after switching the ignition on; it should then go out. If the light stays on when riding over 10 mph, stop the machine and recheck the pressure in the system. If the light fails to illuminate, suspect bulb failure before checking the warning system. If the light illuminates but fails to go out, the warning system should be checked. Refer to Section 22 for details of bulb removal.
2   The warning light control unit is mounted within the dummy fuel tank, on the left-hand side, just beneath the fuse box. Should the unit need to be renewed, it may be unplugged from the wiring harness by detaching the block connector and then pulled up out of its rubber mounting casing. Note that it is not possible to repair this unit due to its being a sealed item.
3   The pressure sensor switch for the warning system is located behind the right-hand sidepanel and forms part of the air charging point assembly. The switch can only be unscrewed from the 3-way charging union after pressure has been released from the system and the union detached from its mounting plate and pulled clear of the machine. Full details of union removal and refitting are contained in Section 10 of Chapter 5.
4   Before testing the system components check that the air pressure in the rear suspension units is within the specified limits; check the fuse for the system has not blown or is defective; check that the warning light bulb has not failed and finally, check all system wires and wiring connections for continuity and signs of chafing, damage, corrosion or looseness. If the above checks prove satisfactory, set a multimeter to the resistance function and check the sensor switch for continuity. With the pressure in the suspension units above 46 psi (2.6 kg/sq cm), there should be continuity. No continuity should register once the system pressure falls below 28 psi (2.0 kg/sq cm). If the sensor switch is found to be serviceable, suspect the control unit.

15 Switch testing procedure: general

1   In the event of a suspected malfunction the various switch contacts can be checked for continuity in a similar manner to that described in the two preceding Sections. Details of the switch connections and the appropriate wiring will be found in the colour wiring diagrams that follow. Note that the electrical system should be isolated by disconnecting the battery leads to avoid short circuits.

10.2 Slacken the throttle cable adjusters to allow removal of the switch

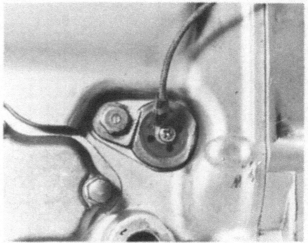

12.1 The neutral indicator switch is located in the right-hand wall of the crankcase

14.1 The rear suspension air pressure warning light control unit is contained within a rubber casing

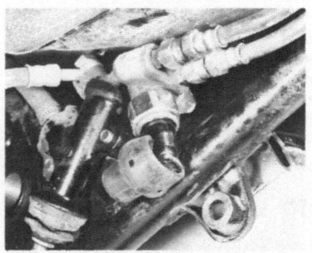

14.3 The air pressure sensor switch forms part of the charging point union

16 Headlamp: bulb renewal

1 The headlamp bulb is held in a separate bulb holder which fits into the rear of the reflector. Access to the holder is gained by detaching the headlamp rim, complete with glass and reflector. On Standard models, remove the rim assembly by unscrewing the two screws which pass through the shell in the 4 o'clock and 8 o'clock positions. Pull the rim out at the bottom and then away from the shell. On Interstate models, first remove the beam adjusting knob from within the fairing by loosening the set screw and pulling the knob off the shaft. Unscrew and remove the nut from the threaded sleeve and displace the spring washer and plain washer. The headlamp assembly can now be eased out of the fairing.

2 After displacing the headlamp unit pull the socket from the bulb pins at the rear of the holder. Prise off the rubber boot which protects the bulb holder. The bulb is secured by two sprung arms which hinge from one side of the holder. Pinch the arms together to free them and then lift the bulb out. The bulb is of the tungsten-halogen type with a quartz envelope. It is important that the envelope is not touched with the fingers, because any greasy or acidic deposits will etch the quartz leading to the early failure of the bulb. If the envelope is touched

inadvertently, it should be cleaned with a solvent such as methylated spirits. On UK models, the pilot bulb holder is a press fit in the reflector, the bulb being of the bayonet fitting type. To remove the bulb, press it inwards and turn it anti-clockwise so that the bayonet pins disengage.

17 Headlamp: adjusting beam height – Standard models

1 Beam height is adjusted by rotating the headlamp about the two headlamp shell retaining bolts, after these have been slackened off. Horizontal adjustment is provided by a screw which passes into the headlamp rim.

2 UK lighting regulations stipulate that the lighting system must be arranged so that the light will not dazzle a person standing at a distance greater than 25 feet from the lamp, whose eye level is not less than 3 ft 6 inches above that plane. It is easy to approximate this setting by placing the machine 25 feet away from a wall, on a level road and setting the dipped beam height so that it is concentrated at the same height as the distance of the centre of the headlamp from the ground. The rider must be seated normally during this operation and also the pillion passenger, if one is carried regularly.

16.1a Remove the headlamp rim retaining screws

16.1b Detach the socket from the rear of the reflector unit

16.2a Release the spring clips and remove the headlamp bulb

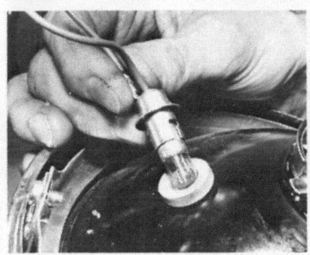

16.2b The pilot bulb holder is a press fit on the reflector (UK only)

17.1 Headlamp beam horizontal adjustment is provided by a screw passing through the headlamp rim

18 Headlamp: adjusting beam height – Interstate models

1   The headlamp mounted in the fairing of the Interstate model may be adjusted for beam height by turning the knob located within the fairing behind the headlamp. Horizontal adjustment is provided by a screw set into the headlamp rim at the 8 o'clock position.

2   Refer to the preceding Section for information on setting of the headlamp beam.

19 Tail/stop lamp: bulb renewal

1   To gain access to the two twin-filament bulbs mounted within the rear lamp unit, unscrew and remove the four crosshead screws that retain the plastic lens in position. Separate the lens from the lamp unit, taking care to ensure that the rubber seal is retained in position on the case rim and does not tear or split.

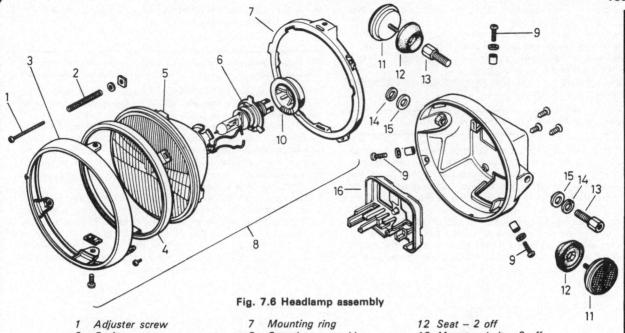

Fig. 7.6 Headlamp assembly

1 Adjuster screw	7 Mounting ring	12 Seat – 2 off
2 Spring	8 Complete assembly	13 Mountng bolt – 2 off
3 Outer rim	9 Screw – 3 off	14 Spring washer – 2 off
4 Inner rim	10 Boot	15 Plain washer – 2 off
5 Reflector unit	11 Reflector – 2 off	16 Connector holder
6 Bulb		

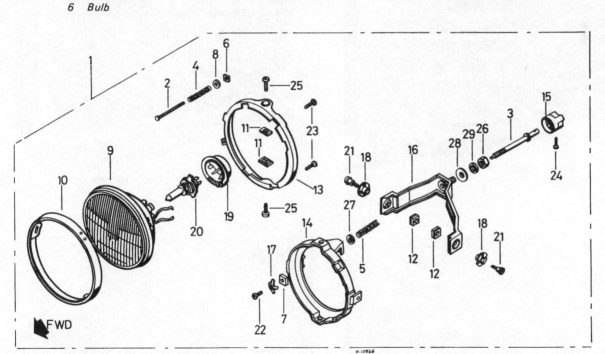

Fig. 7.7 Headlamp assembly – Interstate model

1 Complete assembly	11 Spring nut – 2 off	21 Mounting bolt – 2 off
2 Adjuster screw	12 Nut – 2 off	22 Screw
3 Adjuster spindle	13 Mounting ring	23 Screw – 2 off
4 Spring	14 Mounting ring	24 Screw
5 Spring	15 Adjuster knob	25 Screw – 2 off
6 Special nut	16 Mounting bracket	26 Nut
7 Special nut	17 Spring washer	27 Washer
8 Washer	18 Special washer – 2 off	28 Washer
9 Reflector unit	19 Boot	29 Spring washer
10 Headlamp rim	20 Bulb	

2   The bulbs have a bayonet fitting with offset pins so that they can be fitted in one position only. Take care not to touch the glass envelope of the new bulb when fitting, use a dry, clean cloth or tissue. To remove a bulb, push it inwards and turn it anti-clockwise so that the bayonet pins disengage. Fitting a bulb is the reversal of this procedure.

3   When refitting the lens, take care not to over-tighten the screws and crack the plastic.

20 Direction indicator lamps: bulb renewal

1   Direction indicator lamps are fitted to the front and rear of the machine. To replace a bulb, remove the plastic lens cover by withdrawing the three retaining crosshead screws. Push the bulb in, turn it to the left and withdraw. Note it is important to use a correctly rated bulb otherwise the flashing rate will be altered.

2   Note that all rear fitted bulbs are of the single-filament type as are the front bulbs of Standard models supplied to the UK market. Interstate models and Standard models supplied to the USA are fitted with double-filament bulbs with offset pins to prevent incorrect repositioning on replacement. The second

filament is separately switched to allow the front indicators to serve as daylight running lamps.

21 Direction indicator relay unit: location and replacement

1   The direction indicator relay unit is located within the dummy fuel tank, on the left-hand side, and is supported on an anti-vibration mounting made of rubber.

2   If the flasher unit is functioning correctly, a series of audible clicks will be heard when the indicator lamps are in action. If the unit malfunctions and all the bulbs are in working order, the usual symptom is one initial flash before the unit goes dead; it will be necessary to replace the unit complete if the fault cannot be attributed to any other cause.

3   Remove the unit by noting the position of the electrical leads connected to the unit before disconnecting them and pulling the unit down out of its rubber holder. The new unit may be fitted by reversing this procedure. Care should be taken when handling the unit to ensure that it is not dropped or otherwise subjected to shocks. The internal components may be irreparably damaged by such treatment.

19.1a Remove the tail lamp lens ...

19.1b ... to gain access to the two twin-filament bulbs

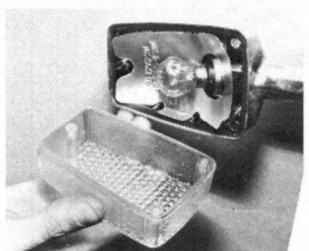

20.1 Remove the indicator lens to expose the bulb

21.1 The direction indicator relay unit is mounted within the dummy fuel tank

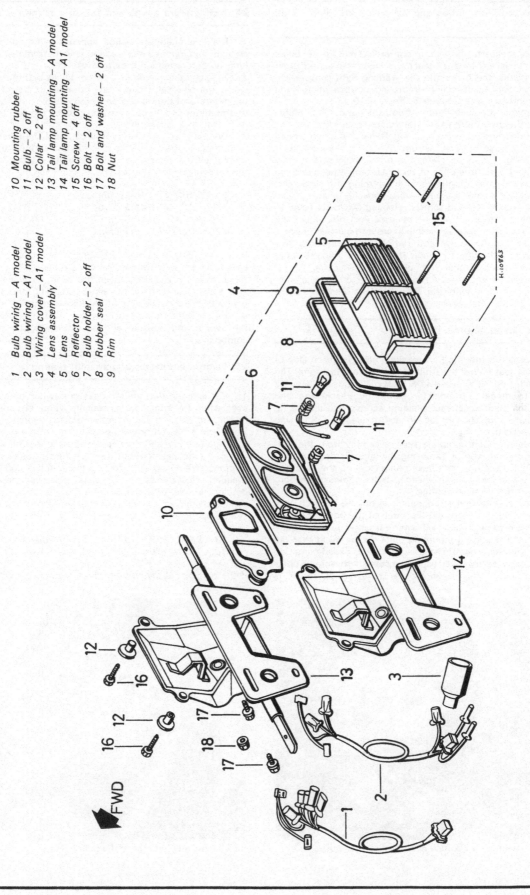

Fig. 7.8 Tail lamp

1 Bulb wiring – A model
2 Bulb wiring – A1 model
3 Wiring cover – A1 model
4 Lens assembly
5 Lens
6 Reflector
7 Bulb holder – 2 off
8 Rubber seal
9 Rim
10 Mounting rubber
11 Bulb – 2 off
12 Collar – 2 off
13 Tail lamp mounting – A model
14 Tail lamp mounting – A1 model
15 Screw – 4 off
16 Bolt – 2 off
17 Bolt and washer – 2 off
18 Nut

H·10963

FWD

22 Instrument illumination and warning panel lights: bulb renewal

1   To gain access to the instrument and warning panel bulbs it will be necessary to disconnect the instruments from their mounting plates and dismantle the warning light console. On Interstate models access will be improved for dismantling if the fairing is removed. See Chapter 5, Section 19.

2   To remove the instruments from their mounting plates disconnect the drive cables at the instrument heads by unscrewing the knurled ring on each cable end. Remove the two domed nuts from the base of each instrument and lift the instrument head out of the mounting plate. The rubber bulb holders may now be pulled out of the base of each instrument head and the bulbs released by pressing inwards and turning anti-clockwise.

3   To dismantle the warning light console, unscrew and remove the four crosshead screws retaining the lower cover to the console assembly. It will be necessary to disconnect the headlamp shell from its pivot points in order to gain easy access to these screws. Unscrew and remove the remaining two screws securing the top of the unit and lift away both the upper and lower covers to expose the bulb holders.

4   Before renewing warning light bulbs, check the recommended wattage with the Specifications Section at the beginning of this Chapter.

23 Horns: location and adjustment

1   The horns are mounted on brackets attached each side of the machine just below the front of the dummy fuel tank. The rear of each horn is encased in a plastic shroud; this may be removed to reveal the adjuster screw by unscrewing and removing the two crosshead retaining screws and unclipping the two halves of the shroud so as to release it from the mounting bracket.

2   Removal of each complete horn assembly from the machine may be achieved by removing the shroud, disconnecting the electrical leads from the push connectors on the horn and unscrewing the two hexagon-headed bolts retaining the horn assembly to the frame mounting.

3   A screw and locknut arrangement at the rear of each horn provides a means of adjusting the pitch of the note produced. Clear access may be gained to this central screw by pivoting the horn clear of the shroud mounting plate. In the event that the horn's performance deteriorates significantly, experimentation with the screw setting will usually restore it to normal operation. If a horn fails, it should be renewed as repair is impracticable.

24 Temperature gauge and sensor: testing

1   If the temperature gauge appears to be faulty, drain the coolant and remove the sensor switch from the thermostat housing before testing it as follows.

2   Suspend the switch in a pan of oil so that the sensor tip is below the oil level. Place a thermometer in the oil so that the temperature of the oil can be noted. Neither the switch or the thermometer should be allowed to touch the pan because this will result in a false reading.

3   Wearing eye and skin protection, heat the oil slowly. Set a multimeter to the resistance function and connect the probes to the switch terminals. Note the resistance readings obtained at the following temperatures.

     60°C (140°F)   104.0 ohms
     85°C (185°F)   43.9 ohms
     110°C (232°F) 20.3 ohms
     120°C (250°F) 16.1 ohms

If the readings obtained differ from those listed above, the sensor switch is faulty and must be renewed. If the switch is found to be serviceable, check the wiring between the gauge and sensor for continuity before returning the temperature gauge to a Honda Service Agent for examination and renewal.

4   Before refitting the sensor switch, clean the threads of both the switch and housing and coat them lightly with sealing compound.

25 Electric fan motor and switch: testing

1   If it is found that the fan does not switch on automatically even when the engine is excessively warm, disconnect the fan motor and test it independently from the switch.

2   Using a 12 volt power source, connect the positive lead to the Blue terminal and the negative lead to the Black terminal. If th motor functions correctly, remove the sensor switch from the thermostat housing after having lowered the coolant level by draining. Test the switch by using a similar method to that detailed in the previous Section for the temperature gauge sensor switch and check that the temperatures at which the switch opens and closes correspond with those given below.

     93 - 97°C (199 - 207°F)   Switch open (no continuity)
     98 - 102°C (208 - 215°F) Switch closed (continuity)

If the readings obtained differ, then the sensor switch must be renewed.

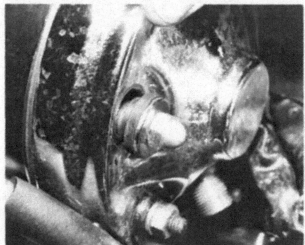

22.2a Remove the two domed nuts from the base of each instrument ...

22.2b ... to allow access to the bulb holders

22.3a Remove the lower console cover retaining screws

22.3b The warning panel lights are mounted in a rubber covered block

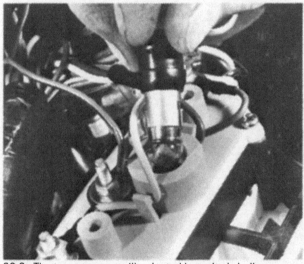
22.3c The two gauges are illuminated by a single bulb

23.1a Release the horn shroud by removing the two screws ...

23.1b ... and unclipping it from the mounting bracket

23.3 Pivot the horn clear of the shroud mounting plate to gain access to the central adjuster screw

26 Fuel gauge and float switch: testing

1   Because the fuel is contained in a tank below the dualseat and therefore verification of the amount of fuel is difficult, a fuel gauge is incorporated in the system.

2   If the fuel gauge malfunctions, carry out the following test to isolate where the trouble lies. Remove the dualseat. Pull the two leads off the float switch situation in top of the tank. Turn on the ignition and touch the two leads together. The needle on the fuel gauge should deflect to the 'full' position. If the gauge functions satisfactorily, suspect the float switch.

3   The float switch can be removed from the tank by turning the switch retaining ring anti-clockwise so as to release it from the spigots on the tank housing. Honda recommend that special tool No 006106 is used for this purpose but it was found that inserting the nose ends of a pair of long-nose pliers into the ring slots and turning the pliers served to rotate the ring. When refitting the float switch, ensure that the arrow marked on the ring aligns with the arrow on the tank housing once it is fully tightened in position.

26.3a Remove the switch retaining ring by turning it anti-clockwise ...

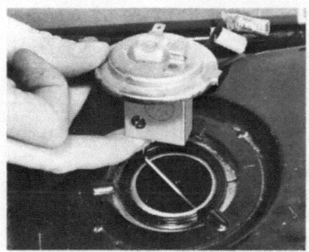

26.3b ... to allow the switch to be lifted clear of the tank

26.3c Align the arrow on the retaining ring with that on the tank housing

27 Fault diagnosis: electrical system

Symptom	Cause	Remedy
Complete electrical failure	Blown fuse(s)	Check wiring and electrical components for short circuit before fitting new fuse.
	Isolated battery	Check battery connections, also whether connections show signs of corrosion.
Dim lights, horn and starter inoperative	Discharged battery	Remove battery and charge with battery charger. Check generator output and voltage regulator rectifier condition.
Constantly blowing bulbs	Vibration or poor earth connection	Check security of bulb holders. Check earth return connections.
Parking lights dim rapidly	Battery will not hold charge	Renew battery at earliest opportunity.
Tail lamp fails	Blown bulb or fuse	Renew.
Headlamp fails	Blown bulb or fuse	Renew.
Flashing indicators do not operate, or flash fast or slow	Blown bulb Damaged flasher unit	Renew bulb. Renew flasher unit.
Horn inoperative or weak	Faulty switch Incorrect adjustment	Check switch. Adjust.
Incorrect charging	Faulty alternator Faulty rectifier Faulty regulator Wiring fault	Check. Check. Check and adjust. Check.
Over or under charging	As above, or faulty battery	Check.
Starter motor sluggish	Worn brushes Dirty commutator	Remove starter motor and renew brushes. Clean.
Starter motor does not turn	Machine in gear Emergency switch in OFF position Faulty switches or wiring Battery flat Loose battery terminal connection(s)	Disengage clutch. Turn on. Check continuity. Recharge. Check and tighten if necessary.

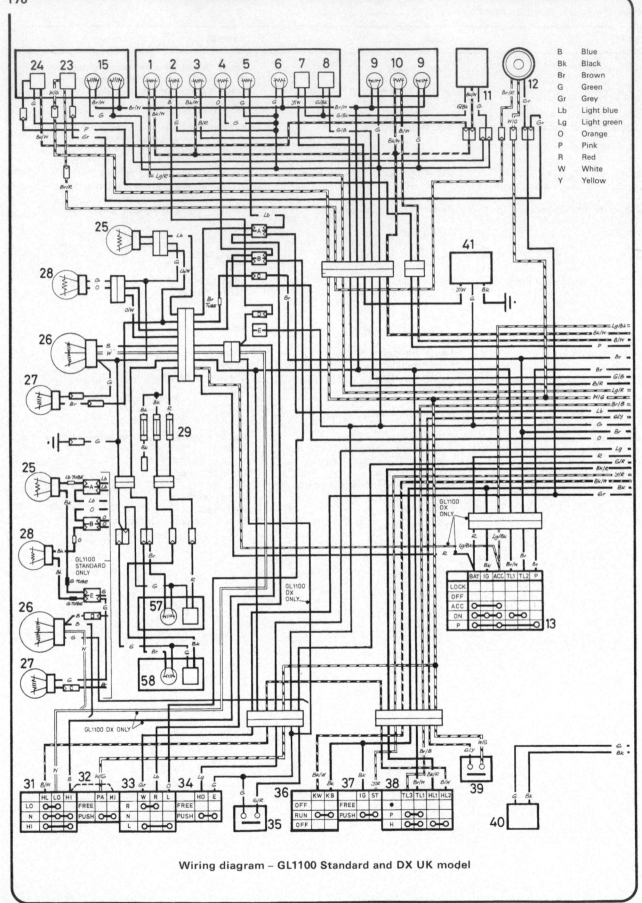

B	Blue		
Bk	Black		
Br	Brown		
G	Green		
Gr	Grey		
Lb	Light blue		
Lg	Light green		
O	Orange		
P	Pink		
R	Red		
W	White		
Y	Yellow		

Wiring diagram – GL1100 Standard and DX UK model

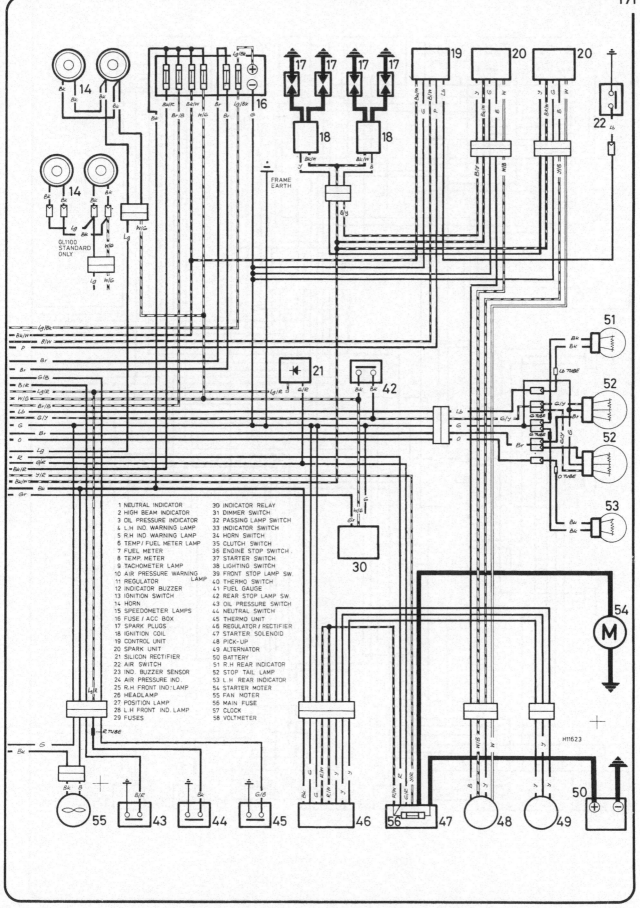

1 NEUTRAL INDICATOR
2 HIGH BEAM INDICATOR
3 OIL PRESSURE INDICATOR
4 L.H. IND. WARNING LAMP
5 R.H. IND. WARNING LAMP
6 TEMP/FUEL METER LAMP
7 FUEL METER
8 TEMP. METER
9 TACHOMETER LAMP
10 AIR PRESSURE WARNING LAMP
11 REGULATOR
12 INDICATOR BUZZER
13 IGNITION SWITCH
14 HORN
15 SPEEDOMETER LAMPS
16 FUSE / ACC BOX
17 SPARK PLUGS
18 IGNITION COIL
19 CONTROL UNIT
20 SPARK UNIT
21 SILICON RECTIFIER
22 AIR SWITCH
23 IND. BUZZER SENSOR
24 AIR PRESSURE IND.
25 R.H FRONT IND. LAMP
26 HEADLAMP
27 POSITION LAMP
28 L.H. FRONT IND. LAMP
29 FUSES

30 INDICATOR RELAY
31 DIMMER SWITCH
32 PASSING LAMP SWITCH
33 INDICATOR SWITCH
34 HORN SWITCH
35 CLUTCH SWITCH
36 ENGINE STOP SWITCH
37 STARTER SWITCH
38 LIGHTING SWITCH
39 FRONT STOP LAMP SW.
40 THERMO SWITCH
41 FUEL GAUGE
42 REAR STOP LAMP SW.
43 OIL PRESSURE SWITCH
44 NEUTRAL SWITCH
45 THERMO UNIT
46 REGULATOR / RECTIFIER
47 STARTER SOLENOID
48 PICK-UP
49 ALTERNATOR
50 BATTERY
51 R.H REAR INDICATOR
52 STOP TAIL LAMP
53 L.H. REAR INDICATOR
54 STARTER MOTER
55 FAN MOTER
56 MAIN FUSE
57 CLOCK
58 VOLTMETER

GL1100 STANDARD ONLY

FRAME EARTH

H11623

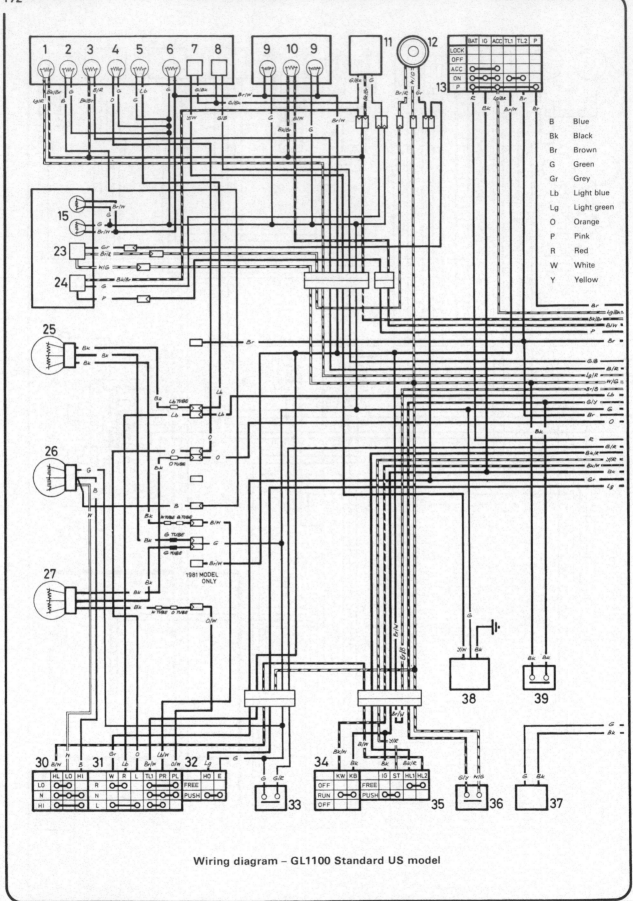

Wiring diagram – GL1100 Standard US model

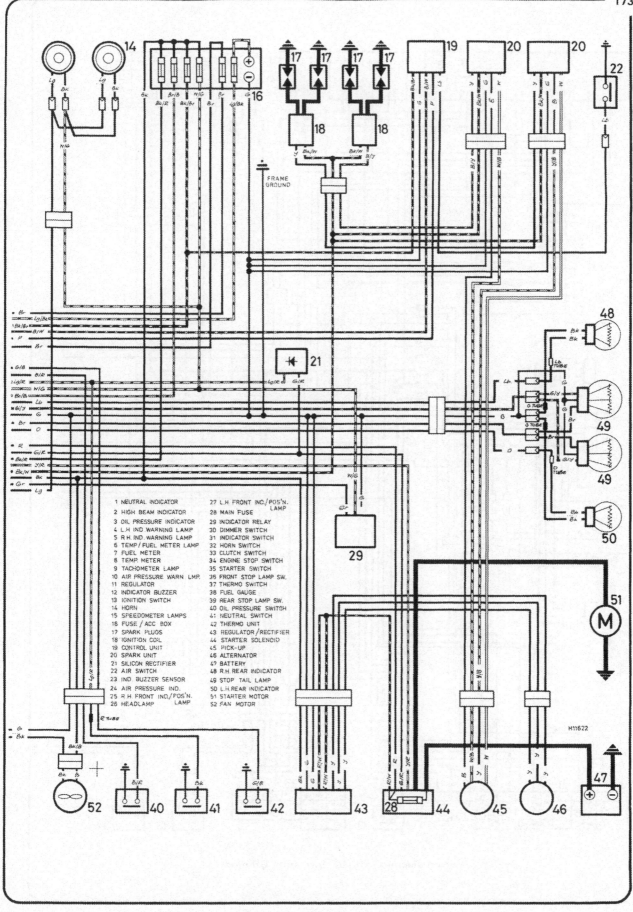

1 NEUTRAL INDICATOR
2 HIGH BEAM INDICATOR
3 OIL PRESSURE INDICATOR
4 L.H. IND. WARNING LAMP
5 R.H. IND. WARNING LAMP
6 TEMP./FUEL METER LAMP
7 FUEL METER
8 TEMP. METER
9 TACHOMETER LAMP
10 AIR PRESSURE WARN. LMP.
11 REGULATOR
12 INDICATOR BUZZER
13 IGNITION SWITCH
14 HORN
15 SPEEDOMETER LAMPS
16 FUSE / ACC BOX
17 SPARK PLUGS
18 IGNITION COIL
19 CONTROL UNIT
20 SPARK UNIT
21 SILICON RECTIFIER
22 AIR SWITCH
23 IND. BUZZER SENSOR
24 AIR PRESSURE IND.
25 R.H. FRONT IND./POS'N. LAMP
26 HEADLAMP

27 L.H. FRONT INC./POS'N. LAMP
28 MAIN FUSE
29 INDICATOR RELAY
30 DIMMER SWITCH
31 INDICATOR SWITCH
32 HORN SWITCH
33 CLUTCH SWITCH
34 ENGINE STOP SWITCH
35 STARTER SWITCH
36 FRONT STOP LAMP SW.
37 THERMO SWITCH
38 FUEL GAUGE
39 REAR STOP LAMP SW.
40 OIL PRESSURE SWITCH
41 NEUTRAL SWITCH
42 THERMO UNIT
43 REGULATOR/RECTIFIER
44 STARTER SOLENOID
45 PICK-UP
46 ALTERNATOR
47 BATTERY
48 R.H. REAR INDICATOR
49 STOP TAIL LAMP
50 L.H. REAR INDICATOR
51 STARTER MOTOR
52 FAN MOTOR

FRAME GROUND

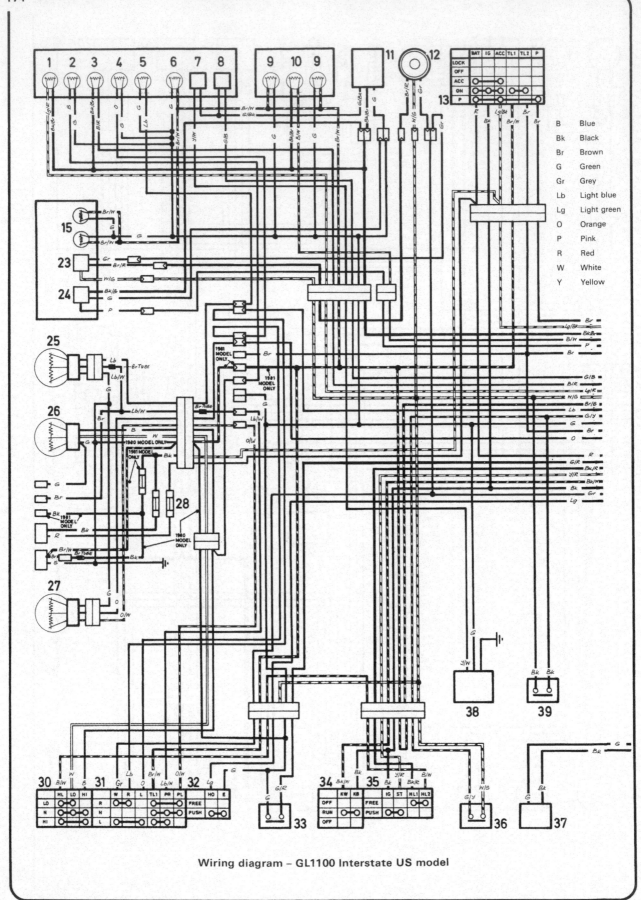

Wiring diagram – GL1100 Interstate US model

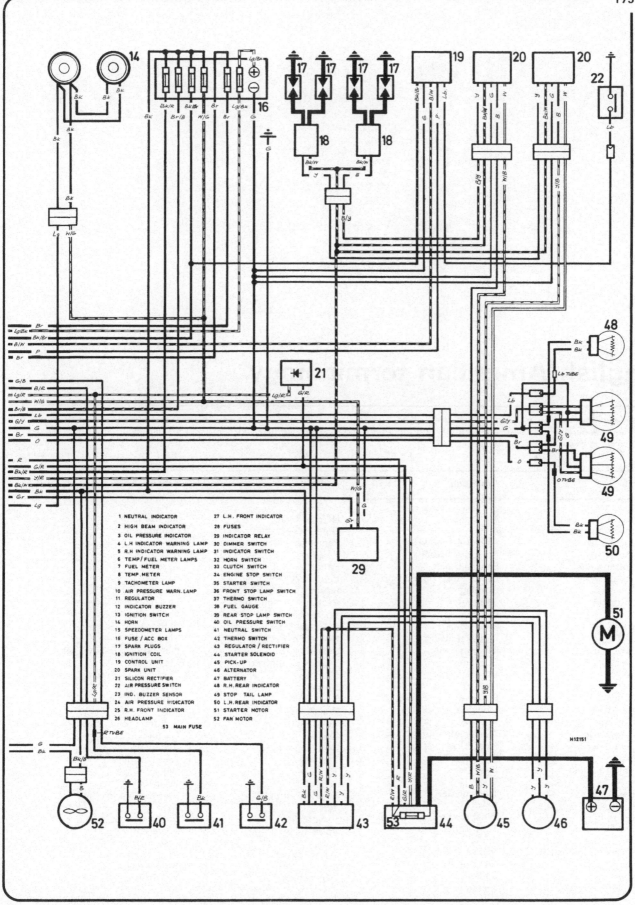

1 NEUTRAL INDICATOR
2 HIGH BEAM INDICATOR
3 OIL PRESSURE INDICATOR
4 L.H INDICATOR WARNING LAMP
5 R.H. INDICATOR WARNING LAMP
6 TEMP/FUEL METER LAMPS
7 FUEL METER
8 TEMP. METER
9 TACHOMETER LAMP
10 AIR PRESSURE WARN. LAMP
11 REGULATOR
12 INDICATOR BUZZER
13 IGNITION SWITCH
14 HORN
15 SPEEDOMETER LAMPS
16 FUSE / ACC BOX
17 SPARK PLUGS
18 IGNITION COIL
19 CONTROL UNIT
20 SPARK UNIT
21 SILICON RECTIFIER
22 AIR PRESSURE SWITCH
23 IND. BUZZER SENSOR
24 AIR PRESSURE INDICATOR
25 R.H. FRONT INDICATOR
26 HEADLAMP

27 L.H. FRONT INDICATOR
28 FUSES
29 INDICATOR RELAY
30 DIMMER SWITCH
31 INDICATOR SWITCH
32 HORN SWITCH
33 CLUTCH SWITCH
34 ENGINE STOP SWITCH
35 STARTER SWITCH
36 FRONT STOP LAMP SWITCH
37 THERMO SWITCH
38 FUEL GAUGE
39 REAR STOP LAMP SWITCH
40 OIL PRESSURE SWITCH
41 NEUTRAL SWITCH
42 THERMO SWITCH
43 REGULATOR / RECTIFIER
44 STARTER SOLENOID
45 PICK-UP
46 ALTERNATOR
47 BATTERY
48 R.H. REAR INDICATOR
49 STOP TAIL LAMP
50 L.H. REAR INDICATOR
51 STARTER MOTOR
52 FAN MOTOR

53 MAIN FUSE

H12151

English/American terminology

Because this book has been written in England, British English component names, phrases and spellings have been used throughout. American English usage is quite often different and whereas normally no confusion should occur, a list of equivalent terminology is given below.

English	American	English	American
Air filter	Air cleaner	Number plate	License plate
Alignment (headlamp)	Aim	Output or layshaft	Countershaft
Allen screw/key	Socket screw/wrench	Panniers	Side cases
Anticlockwise	Counterclockwise	Paraffin	Kerosene
Bottom/top gear	Low/high gear	Petrol	Gasoline
Bottom/top yoke	Bottom/top triple clamp	Petrol/fuel tank	Gas tank
Bush	Bushing	Pinking	Pinging
Carburettor	Carburetor	Rear suspension unit	Rear shock absorber
Catch	Latch	Rocker cover	Valve cover
Circlip	Snap ring	Selector	Shifter
Clutch drum	Clutch housing	Self-locking pliers	Vise-grips
Dip switch	Dimmer switch	Side or parking lamp	Parking or auxiliary light
Disulphide	Disulfide	Side or prop stand	Kick stand
Dynamo	DC generator	Silencer	Muffler
Earth	Ground	Spanner	Wrench
End float	End play	Split pin	Cotter pin
Engineer's blue	Machinist's dye	Stanchion	Tube
Exhaust pipe	Header	Sulphuric	Sulfuric
Fault diagnosis	Trouble shooting	Sump	Oil pan
Float chamber	Float bowl	Swinging arm	Swingarm
Footrest	Footpeg	Tab washer	Lock washer
Fuel/petrol tap	Petcock	Top box	Trunk
Gaiter	Boot	Torch	Flashlight
Gearbox	Transmission	Two/four stroke	Two/four cycle
Gearchange	Shift	Tyre	Tire
Gudgeon pin	Wrist/piston pin	Valve collar	Valve retainer
Indicator	Turn signal	Valve collets	Valve cotters
Inlet	Intake	Vice	Vise
Input shaft or mainshaft	Mainshaft	Wheel spindle	Axle
Kickstart	Kickstarter	White spirit	Stoddard solvent
Lower leg	Slider	Windscreen	Windshield
Mudguard	Fender		

Index